Mobile User Research

A Practical Guide

Synthesis Lectures on Mobile and Pervasive Computing

Editor

Mahadev Satyanarayanan, *Carnegie Mellon University*

Synthesis Lectures on Mobile and Pervasive Computing is edited by Mahadev Satyanarayanan of Carnegie Mellon University. Mobile computing and pervasive computing represent major evolutionary steps in distributed systems, a line of research and development that dates back to the mid-1970s. Although many basic principles of distributed system design continue to apply, four key constraints of mobility have forced the development of specialized techniques. These include: unpredictable variation in network quality, lowered trust and robustness of mobile elements, limitations on local resources imposed by weight and size constraints, and concern for battery power consumption. Beyond mobile computing lies pervasive (or ubiquitous) computing, whose essence is the creation of environments saturated with computing and communication, yet gracefully integrated with human users. A rich collection of topics lies at the intersections of mobile and pervasive computing with many other areas of computer science.

Mobile User Research: A Practical Guide
Sunny Consolvo, Frank R. Bentley, Eric B. Hekler, and Sayali S. Phatak
May 2017

Pervasive Displays: Understanding the Future of Digital Signage No Access
Nigel Davies, Sarah Clinch, and Florian Alt
April 2014

Cyber Foraging: Bridging Mobile and Cloud Computing
Jason Flinn
September 2012

Mobile Platforms and Development Environments No Access
Sumi Helal, Raja Bose, Wendong Li
February 2012

Quality of Service in Wireless Networks Over Unlicensed Spectrum No Access
Klara Nahrstedt
November 2011

The Landscape of Pervasive Computing Standards No Access
Sumi Helal
June 2010

A Practical Guide to Testing Wireless Smartphone Applications No Access
Julian Harty
2009

Location Systems: An Introduction to the Technology Behind Location Awareness No Access
Anthony LaMarca and Eyal de Lara
2008

Replicated Data Management for Mobile Computing No Access
Douglas B. Terry
2008

Application Design for Wearable Computing No Access
Dan Siewiorek, Asim Smailagic, and Thad Starner
2008

Controlling Energy Demand in Mobile Computing Systems No Access
Carla Schlatter Ellis
2007

RFID Explained: A Primer on Radio Frequency Identification Technologies No Access
Roy Want
2006

Mobile User Research: A Practical Guide
Sunny Consolvo, Frank R. Bentley, Eric B. Hekler, and Sayali S. Phatak

ISBN: 978-3-031-01357-7 print
ISBN: 978-3-031-02485-6 ebook

DOI: 10.1007/978-031-02485-6

A Publication in the Springer series
SYNTHESIS LECTURES ON MOBILE AND PERVASIVE COMPUTING, #12
Series Editor: Mahadev Satyanarayanan, Carnegie Mellon University

Series ISSN: 1933-9011 Print   1933-902X Electronic

Mobile User Research

A Practical Guide

Sunny Consolvo
Google

Frank R. Bentley
Yahoo

Eric B. Hekler
Arizona State University

Sayali S. Phatak
Arizona State University

SYNTHESIS LECTURES ON MOBILE AND PERVASIVE COMPUTING #12

ABSTRACT

This book will give you a practical overview of several methods and approaches for designing mobile technologies and conducting mobile user research, including how to understand behavior and evaluate how such technologies are being (or may be) used out in the world. Each chapter includes case studies from our own work and highlights advantages, limitations, and very practical steps that should be taken to increase the validity of the studies you conduct and the data you collect.

This book is intended as a practical guide for conducting mobile research focused on the user and their experience. We hope that the depth and breadth of case studies presented, as well as specific best practices, will help you to design the best technologies possible and choose appropriate methods to gather ethical, reliable, and generalizable data to explore the use of mobile technologies out in the world.

KEYWORDS

mobile computing, user research, qualitative research, quantitative research, theory, experimental design, sensors, field study, lab study, usability, analytics, mHealth, digital health, behavior change technologies

Contents

Preface xiii

Acknowledgments xv

1 Introduction to Mobile User Research **1**
    1.1   User Study Basics 3
        1.1.1   User Study Methods 3
        1.1.2   Data Analysis 6
        1.1.3   Recruiting Participants 8
        1.1.4   Participant Incentives 9
        1.1.5   Field Study Logistics 10
        1.1.6   Ethics, Consent, and Review 11
        1.1.7   Pilot and Review Everything 12

2 Sensor and Usage Data **15**
    2.1   Data Types 16
    2.2   General Uses of These Data 22
        2.2.1   Understanding Overall Device Use 26
    2.3   Factors to Take into Account When Selecting Data Sources 30
        2.3.1   Practical Suggestions 36
    2.4   Conclusion 39

3 Observations in the Field and in the Lab **41**
    3.1   Introduction 41
    3.2   Exploratory Field Studies 42
        3.2.1   Field Study Tips 43
        3.2.2   Examples of Generative Field Studies 47
    3.3   Evaluative Field Studies 52
        3.3.1   Preparing for an Evaluative Field Study 52
        3.3.2   Wizard of Oz Methods 55
    3.4   Lab Usability Studies 60
        3.4.1   Limitations 61
        3.4.2   Lab Usability Study Tips 62
    3.5   Summary 69

4   Diary Studies and Experience Sampling **71**
    4.1   Introduction 71
    4.2   Diary Studies 71
        4.2.1   Basic Method 73
        4.2.2   Variations 76
        4.2.3   Limitations 78
        4.2.4   Case Studies 79
    4.3   The Experience Sampling Method 83
        4.3.1   History 84
        4.3.2   Basic Method 87
        4.3.3   Limitations 91
        4.3.4   Case Studies 92
    4.4   Summary 99

5   Answering "Did it work?":
A Primer to Experimental Designs to Test for Change **101**
    5.1   Establishing Cause and Effect: Science 101 102
    5.2   Primer of Experimental Designs 105
        5.2.1   Within-Person Quasi-Experimental Designs 110
        5.2.2   Between-Person Quasi-Experimental Designs 115
        5.2.3   Between-Person Experimental Designs 117
        5.2.4   Within-Person Experimental Designs 127
        5.2.5   Other Designs 129
        5.2.6   General Words of Caution 130
    5.3   Summary 131

6   Using Theory in Mobile User Research **133**
    6.1   Introduction 133
    6.2   Defining Terms 134
    6.3   Uses of Behavioral Theory 135
        6.3.1   Understanding the Target Problem: Designing Ways to Observe,
                Measure, and Study 136
        6.3.2   Defining a Target User and Audience 138
        6.3.3   Defining the Design of a Technical System 140
    6.4   Selecting the "Right" Theory(ies) 143
        6.4.1   Using a Familiar Theory 143
        6.4.2   Utilizing User Insights and Previous Research 144
        6.4.3   Meta-model Followed by Conceptual Frameworks 145

6.5   Judging the Quality of a Theory 146

6.6   A Few Theories to Get Started 147

6.7   An Illustrative Case Study: The MILES Study 150

6.8   Summary 153

7   Big Challenges and Open Questions **157**

7.1   Diary Studies and Experience Sampling 157

      7.1.1   Triangulating Data 158

      7.1.2   Evaluating New Experience Sampling Techniques 159

      7.1.3   From Sensors to Usable Information 159

      7.1.4   From "On Average" To Usable Evidence 164

      7.1.5   Empowering End-Users in Personalization of Mobile
              Experiences 165

      7.1.6   From Theories to Computational Models 165

7.2   Summary 166

References **167**

Author Biographies **193**

Preface

This book will give you a practical overview of several methods and approaches for designing mobile technologies and conducting mobile user research, including how to understand behavior and evaluate how such technologies are being (or may be) used out in the world. Each chapter includes case studies from our own work and highlights advantages, limitations, and very practical steps that should be taken to increase the validity of the data you collect.

In Chapter 1, we provide a very brief history of mobile user research. We also include some of our favorite sources and tips for readers who aren't very familiar with the basics of user research in general.

In Chapter 2, we explore many of the rich sensors available on today's mobile devices and the types of information that can be collected from these devices. We also discuss how to instrument applications to capture what people are doing so as to better quantify behavior and improve the design of your system using this data.

In Chapter 3, we discuss exploratory field studies, evaluative field studies, and lab usability studies. *Exploratory* (or *generative*) *field studies* can help you understand the people you're going to design for and the contexts in which they spend their time. *Evaluative field studies* can help you understand how people engage with a technology in the wild. And *lab usability studies* can help you get a design in good shape before you spend the time and effort to go into the field.

Chapter 4 covers diary studies and experience sampling—two of our go-to field study methods. Both are self-report methods that can be used to understand attitudes and behavior over a period of time. With *diary studies*, participants keep a log about their thoughts or experiences, and with *experience sampling*, participants respond to brief questionnaires at specific times or in specific contexts.

We then move to Chapter 5, where we tackle the topic of experimental design. We emphasize how you can answer the seemingly simple question, "Did it work?", when the goal of your technology is to change something about your users' behavior.

In Chapter 6, we offer an accessible guide for how theory can help you. Specifically, we describe how to use theory as a structure for developing and choosing appropriate measures, to better specify and define the target audience for your research, and to help guide the design of your technology.

And finally, we close in Chapter 7 by looking forward. We explore some of the limitations and open challenges when it comes to studying people and their interactions with technology out and about in the world, and we propose open questions for future research.

This book is intended as a practical guide for conducting mobile research focused on the user and their experience. We hope that the depth and breadth of case studies presented, as well as specific best practices, will help you to design the best technologies possible and choose appropriate methods to gather ethical, reliable, and generalizable data to explore the use of mobile technologies out in the world.

Acknowledgments

We would like to extend a very big thank you to the following people:

- our principal reviewer, Mahadev "Satya" Satyanarayanan and our editor from Morgan & Claypool Publishers, Mike Morgan, for the opportunity, feedback, and their incredible patience with us;

- our production team at Morgan & Claypool Publishers, including compositor Deb Gabriel and editorial associate Samantha Draper;

- others who provided helpful reviews, including Dan Russell;

- our many collaborators, without whom the case studies would not have happened, in particular:
    - Abby King
    - Adam Rea
    - Ali Rahimi
    - Anthony LaMarca
    - Beverly Harrison
    - Brett Shelton
    - Chris Beckmann
    - Christian Holz
    - Crysta Metcalf
    - Danny Wyatt
    - David W. McDonald
    - Dirk Haehnel
    - Edward Barrett
    - Gaetano Borriello
    - Gunnar Harboe
    - Ian E. Smith
    - James A. Landay

- Jeff Hightower
- Jason Tabert
- Jeff Towle
- Jon Froehlich
- Jonathan Lester
- Karen Church
- Karl Koscher
- Keith Mosher
- Ken Fishkin
- Louis LeGrand
- Matthew Buman
- Matthai Philipose
- Mike Y. Chen
- Miriam Walker
- Nazanin Andalibi
- Nediyana Daskalova
- Pauline Powledge
- Predrag "Pedja" Klasnja
- Peter Roessler
- Roy Want
- Ryan Libby
- Sandra Winter
- Santosh Basapur
- Tanzeem Choudhury
- Tara Matthews
- Tejaswi Peesapati
- Trevor Pering
- Ying-Yu Chen

- our study sponsors, without whom this book and the case studies would not have happened, including:

  - Google

  - Humana

  - Intel Labs Seattle

  - Motorola Labs

  - Yahoo Labs

  - The National Institutes of Health

- our many participants, who for reasons of privacy, will go unnamed; and

- our many colleagues, reviewers, friends, and family members who provided valuable feedback and support over the years.

CHAPTER 1

Introduction to Mobile User Research

Mobile technologies have dramatically changed the way that people interact with each other and the world. People spend an average of over 3.5 hours per day on their smartphones (Khalaf, 2015), using dozens of different applications in a day. Experts predict that the Internet of Things (IoT) will be thriving in less than 10 years (Anderson and Rainie, 2014). This change in the technology that people use, where and why they use it, and how it's permeating their everyday lives continues to evolve at a rapid pace.

The methods and approaches that we cover in this book will prepare you to understand how people use mobile technologies, how to design technologies that will work in the various contexts in which they'll be used, and how to conduct investigations of these technologies. By their very nature, mobile technologies are meant to be used whenever and wherever we go. As such, they need to be designed and studied in a way that takes their many contexts of use into consideration.

This interest in studying the use of systems out in the world, and in exploring people's experiences with technology has been a major focus of the field of human-computer interaction (HCI). The field of HCI grew from a variety of disciplines more than 40 years ago, from human factors engineers who worked on cockpit design and other mission critical systems of the era, to psychologists interested in creating controlled experiments to understand people's capabilities and reactions to new digital systems, to designers interested in how their designs affected people.

Initial topics in HCI explored people's physical capabilities and the implications of these capabilities on the design of technologies. For example, work such as Fitts' Law (Fitts, 1954; Card et al., 1978) helped to define how and where buttons and text fields should be placed on a computer screen, and how large they should be to make interaction with a mouse more efficient. This focus on efficiency drew from earlier work on mission critical systems as well as computing's early focus on workplace systems and applications.

As HCI evolved, researchers began to explore the tasks that people perform with computing systems—not just in the workplace—and the usability of systems that support these tasks. User studies tended to focus on usability inspection methods (Nielsen and Mack, 1994), often measuring attributes such as learnability, efficiency, memorability, errors, and user satisfaction (Scholtz, 2004). Design patterns emerged to help designers create usable websites, building on established and tested design principles (van Duyne et al., 2007). This era enabled many of today's usable websites and desktop applications that we often take for granted.

With the rise in popularity of the mobile phone and the growing area of ubiquitous computing (Weiser, 1991), a new wave of HCI research emerged that was much more interested in how technology was permeating everyday life. Technology was becoming increasingly personal and pervasive; people were using technology not because they had to in order to complete a work task, but because they wanted to—and they were doing it in a variety of places. This shift in technology and its use required a shift in how technology was studied and designed (Abowd and Mynatt, 2000).

This shift was not without its challenges. Whereas most web interactions or desktop applications could reasonably be studied from a fixed place such as a lab or workplace, studying the use of mobile technology required methods where experiences could be observed out in the world, in natural settings. However, the small size of mobile screens and the pervasive use of mobile devices throughout daily life meant that existing observational methods were not always effective. New ways to capture use in the world, both quantitatively and qualitatively, had to be developed or appropriated to understand how these devices were being used and why they were being used in the way that they were. As a result, researchers started to use methods that had seldom if ever been used in the evaluation of websites. Understanding how various use cases, designs, and interactions were perceived and experienced by people became key to making new technologies that people would choose and continue to use, as did using methods that were more appropriate to study these new attributes and contexts (e.g., with a strong focus on ecological validity) (Scholtz and Consolvo, 2004; Consolvo et al., 2007).

Mobile phones—*smartphones* in particular—by nature of their frequent notifications and ease of switching applications, also provided challenges to the definition of a "task" or certainly the measures of task efficiency and completion that were so frequently studied in earlier waves of HCI research. As people began to use apps in parallel (e.g., finding a restaurant in a local recommendation application, sharing a link to it in a text message to a friend, and then navigating to it in a maps application), interactions became more difficult to study, as a single researcher or company likely did not have direct access to the logs of the complete set of interactions. And while a "task" or use case could be broadly defined as finding a place for dinner and then going there, measuring "success" or "error" became increasingly complex and nuanced. What if the user made it to the restaurant for dinner, but didn't like the food or had trouble finding parking? What if their dinner companion didn't like it? Questions of "success" began to involve a much greater deal of qualitative data and analysis than earlier studies of more task-bounded, work-based systems.

Mobile technologies also enable new ways to study people's lives, even if the focus isn't necessarily on *how* people use technology. For example, mobile phones often include sensors that can sense the user's location, activity (e.g., running, walking, biking, or driving a car), and potentially even their emotions. Phones have access to the user's contacts, calendar, and photo history, and they enable a broad range of communication with others. Companion wearables, such as smartwatches, can measure the user's heart rate or emotional excitation via electrodermal activity. This contextual information can be extremely useful—for example, for the design of a wide variety of applications or simply for understanding what people do every day—but the use of such data has to be carefully weighed. Designing new experiences that leverage rich aspects of a user's context (or the context of those nearby) or collecting that data as part of a study *while* protecting privacy and providing strong security is an active area of research (see, for example, Enev et al., 2011).

1.1   USER STUDY BASICS

Throughout this book, we assume that the reader is familiar with basic concepts of user research, such as how to conduct an interview or usability study, how to analyze data, how to recruit and provide incentives for participants, how to obtain informed consent, and so on. This book aims to provide a practical guide for *mobile* user research, however, it does not teach the basics of user research which are already well-documented elsewhere.

For readers who are less familiar with the basics, in this section, we include some of our favorite sources and tips. We hope that the references below will help you explore these basic topics further, and that this book will help you build on the basics so that you can conduct your own great mobile user research.

1.1.1   USER STUDY METHODS

While this book will cover topics for conducting user research for and with mobile technologies, we encourage you to explore additional resources to develop a strong background in the basics of specific research methods that we reference throughout the book, in particular, semi-structured interviews, usability studies, contextual inquiries, and surveys.

Semi-Structured Interviews

Conducting qualitative interviews is a core skill that is used in many research studies. It is often, but not always, used in conjunction with other study methods. Developing your questions, setting the pace and order of topics, building rapport with participants, and ensuring that you collect valid, reliable data are topics that could each fill their own books. We recommend the following as great introductions to designing interview questions and semi-structured protocols:

- Beebe's *Rapid Assessment Process: An Introduction* (2001)

- Denzin and Lincoln's *The SAGE Handbook of Qualitative Research* (2011)

- Gubrium et al.'s *The SAGE Handbook of Interview Research: The Complexity of the Craft* (2012)

We usually take notes during *and* audio record our interviews (with each participant's consent, of course), then have any audio recordings transcribed for analysis. It's not unusual for a participant to not want to be recorded, so always have a backup plan (e.g., a researcher on hand—in addition to the interviewer—to take detailed notes).

In general, we suggest starting interviews lightly, taking time to learn about your participants and their lives, before delving into deeper, more personal topics. To help ensure high quality data, it's a good idea to ask participants to tell you about specific, recent incidents on topics of interest. For example, instead of asking them what they "usually" do or "how often" they do something, ask about the last time they performed the task in which you're interested in and then about the time before that. Good qualitative research is often semi-structured, meaning that the researcher will follow up on topics of interest and ask additional probing questions to more fully explore what the participant is describing. For example, the participant may mention something you hadn't anticipated that you'll want to learn more about. You'll also want to ask follow-up questions to ensure you know what the participant means rather than assuming you understand what they mean—you'd be surprised by how often follow-up questions reveal that your initial assumptions about what a participant meant were mistaken.

Knowing when to probe deeper, what follow-up questions to ask, and how not to prime participants with a particular response are subtle, but tough-to-learn skills that take practice and critical reflection over time. Getting it right yields rich, insightful data, while getting it wrong compromises the validity of your data, and ultimately your findings.

Usability Studies

It is not unusual for mobile user research studies to involve assessing the usability of a new application, service, or device. Given the nature of mobile computing as something that people use out and about in the world, lab-based usability studies almost always need to be paired with other methods as part of a larger research plan, but they often can be helpful (we discuss their utility further in Chapter 3).

The standards of typical usability studies hold when conducting usability for mobile user research (e.g., having participants "think aloud" (Nielsen, 2012), crafting tasks so as not to bias participants by using words or labels from your interface in the task descriptions, not intervening if the participant gets stuck, keeping a neutral facial expression and voice during the study, and so on). To learn about the basics of conducting usability studies, we recommend getting started with:

- Albert and Tullis' *Measuring the User Experience* (2013)

- Nielsen's *Usability Engineering* (1993)

- Nielsen's *Usability 101: An Introduction to Usability* (2012)

- Nielsen and Mack's *Usability Inspection Methods* (1994)

- Rubin and Chisnell's *Handbook of Usability Testing* (2008)

We often like to capture video of the participant's screen and audio + video of the participant's face and voice (with their permission, of course) during usability studies. We discuss important considerations when capturing video of a participant's screen in Chapter 3 (particularly if they have to authenticate on a device that might reveal their account credentials). When appropriate, eye tracking tools that show you exactly where a participant is looking on screen can be helpful.

Contextual Inquiry

The user's (often changing) context is usually an important factor that can help you understand their actions and choices with technology, or what types of design decisions will likely lead to a solution that will integrate well into their lives. One method you can use to learn about their context is *Contextual Inquiry* in which the researcher goes into the field with the participant to observe their interactions in real contexts of use. This might be in their home, car, workplace, favorite fitness trail, and so on. When conducting a contextual inquiry, it is important to ask participants to perform tasks that are realistic for them (e.g., and not ask them to do something that they haven't done or wouldn't likely do). To learn more about how to conduct a contextual inquiry, we recommend:

- Beyer and Holtzblatt's *Contextual Design* (1998)

Surveys

It is frequently important to understand if a finding that you observed in smaller-scale research (e.g., what you learn from one of the methods mentioned above) is common at a larger scale, or if an area that the research team is focusing on relates to behaviors in the broader population. Surveys can help to answer questions like these. Today, with panels readily available on platforms such as SurveyMonkey, Amazon Mechanical Turk, or Google Consumer Surveys, it can often take just a few hours or days to gather lots of survey data on a topic of interest.

But critical to gathering data that you can trust is careful survey design. Although it can be relatively easy for any researcher to deploy a survey, it is not easy to put together a well-designed survey. Indeed, it is all too common for researchers to quickly generate a survey, deploy it, and get data only to find that the information they gathered from their survey is not helping them with the decisions they are making, or worse, leads them in the wrong direction because they misinterpreted what was learned from a survey.

Before deploying a survey, it is essential for you to clearly think through *why* you think a survey is an appropriate method for *what* you want to learn. This might seem obvious, and it is a critical step in using any research method well, but sadly it is often skipped (perhaps because surveys are so easy to run these days). We suggest that *before* and *during the design* of your survey, you think clearly through the decisions that you expect to make after you collect your survey data as a way to start grounding your thinking (for detailed instructions on how to do this, we suggest Harris (2014). If your decisions are more on the design side (e.g., questions that focus on *how* or *why*) then, more often than not, you'll be better off using more qualitative methods like those discussed above. If your questions are more around issues that focus on things like *how much* or eliciting preferences from an already well-curated list, then a survey may be appropriate.

If you ultimately decide that a survey will get you the information you need to help you make decisions, then the next task involves writing questions that people can actually answer in such a way that the responses they provide will help you with your decision-making. Question and response wording and ordering are extremely important, as is asking questions that people can answer reliably. For example, you may have an idea for a new mobile interaction in mind. Rather than go through the process of designing it, you think that it would be easier if you developed a survey in which you describe your new interaction idea and then asked individuals if they thought it was a good idea. While this might seem like a great idea, often, individuals cannot reliably answer a question like this because they do not have any previous experience with your new interaction. As a general rule of thumb, we often ask ourselves, "would our participants have enough experience with the topic of this question to reliably answer this question?" whenever we design new survey questions. Further, similar to crafting interview questions, focusing on aspects of specific recent behaviors can lead to more reliable data than asking about general frequency of use. To learn more about how to design a good survey, we recommend:

- Fowler's *Improving Survey Questions: Design and Evaluation* (1995)

- Fowler's *Survey Research Methods* (2014)

- Harris' *The Complete Guide to Writing Questionnaires: How to Get Better Information for Better Decisions* (2014)

- Callegaro's *Recent Books and Journals Articles in Public Opinion, Survey Methods, Survey Statistics, and Bit Data* (2017)

1.1.2   DATA ANALYSIS

The methods presented in this book typically generate large amounts of data. This data can be qualitative or quantitative, or a mix of both when seeking to triangulate findings between multiple methods.

Many of the methods we discuss in this book (e.g., interviews and diaries) can generate thousands of individual qualitative quotes from participants that the research team has to make sense of. Methods such as:

- grounded theory (Strauss and Corbin, 1998),

- open coding (Strauss and Corbin, 1990),

- and affinity analysis (see Harboe and Huang, 2015 for a review)

can help to inductively find themes in the data and identify larger patterns across research participants. One approach we often use in our qualitative data analysis (Bentley and Barrett, 2012) is to print individual participant quotes onto post-it notes, then collaboratively work with the rest of the research team to identify themes and broader stories in the data. It's important to note if any of those quotes followed potentially biased probing by the interviewer (even if that biasing was intentional), or if any quotes from an individual participant are repeats or clarifications of earlier quotes. Responses that follow biased questions, or that merely echo the question that the interviewer has asked should be discarded prior to analysis or at least treated appropriately. The best questions are often the most minimal that allow participants to respond on their own, such as "tell me more about that" or "why?" Rapid discount methods perform this type of analysis using researcher's insights (Beyer and Holtzblatt, 1998).

When you're analyzing qualitative data, it's important to remember that rigor is critical, and you shouldn't infer meaning that's not clearly there (that latter point is something with which we see a lot of researchers struggle, especially if they're newer to qualitative methods). When we're advising researchers on how to analyze qualitative data, we often like to ask them to continually ask themselves, *if someone else were to review the data, would they come to the same conclusions? If not, why not?* Also, if multiple members of the research team are analyzing the qualitative data, you'll likely want to assess their *inter-rater reliability* (also called *inter-rater agreement* or *interobserver reliability*), which measures the degree to which multiple raters agree.

In terms of quantitative analyses, there is an almost overwhelming number of data analytic procedures one can use to examine these data. As part of that, there are numerous conventions on how to manage, clean, and organize data. A full review of these approaches is well beyond the scope of this book, but a few useful resources include:

- Grus (2015), which provides a nice review on how to organize and clean data, and also some fundamentals on using Python for doing basic data science;

- Flach (2012), which provides a nice intro to slightly more advanced machine learning;

- Ljung (1998), which is a foundational text on system identification, which is a combined experimental design/analytic approach for modeling complex, dynamic systems; and

- Pearl et al. (2016), which is an excellent primer for understanding causal inference in statistics.

In addition, for a deeper dive on what is possible with various analytic strategies for mobile user research, we suggest the following:

- Hekler et al., 2013a,

- Hekler et al., 2013b,

- Spruijt-Metz et al., 2015,

- Nahum-Shani et al., 2015,

- Hekler et al., 2016a, and

- Patrick et al., 2016.

1.1.3   RECRUITING PARTICIPANTS

Recruiting the right set of participants is one of the most important, and often overlooked, aspects of running any user study (as it can be time consuming, expensive, or both). Whether you're running a qualitative usability study with 12 participants or a large-scale quantitative survey of thousands of participants, your participant population is critical to collecting high quality, trustworthy data. You must carefully consider the attributes of your target user audience that need to be represented by the people who are recruited to participate in your study (rather than, for example, going with a convenience sample of college students and staff from your university or employees from your workplace). In most cases, no matter how closely they match your study participation criteria, it's a good idea to avoid recruiting participants who are in any way affiliated with anyone on the research team.[1] A good rule of thumb is to imagine what you would think if the approach that you'd like to use for recruiting participants in your study were instead being used by other researchers whose reputations were unknown to you—would you trust their results?

Whenever possible, we use professional participant recruiters to find people who match the screener[2] that we've developed for a particular study. You may want to consider what type of representation you need from attributes such as gender, age, highest level of education obtained, income, home location (e.g., urban, suburban, or rural), ownership or use of certain technologies, relationship status, household occupants, and so on. We strongly encourage using a professional recruiter given the time and effort needed to reach representative sets of people you don't know. These firms have access to much broader sets of the population than you can likely find on your own. Depending on the type of study, a professional recruiter often costs between $US 75–150 per participant as

---

[1]   We often consider pilot studies to be an exception to this strong suggestion.

[2]   A screener is an instrument (often a survey or structured interview script) that recruiters use to determine if a participant matches the criteria for participation that you have established for your study.

of this printing.[3] If using a professional recruiter is not an option, wide postings across Craigslist, Facebook, Twitter, and grocery store, library, or church notice boards can also help to reach a wider set of potential participants. It is important to reach beyond your own social networks and bubbles if you do turn to social media. Looking at occupations, income, and education levels can help you to know when you're getting a broader set of participants. It's usually a good idea to over-recruit by at least a few participants to account for no-shows and drop-outs (e.g., if you want to ensure that your study includes *at least* 10 participants, you should probably recruit and have the resources to support 12–13 participants).

Panels for online study platforms such as SurveyMonkey, Amazon Mechanical Turk, or Google Consumer Surveys can fluctuate—sometimes they'll be great for what you'll need, while other times, they won't. It's best to compare your study's needs to what the platforms currently support to see if it's a good match or if you need to find a different platform provider.

1.1.4 PARTICIPANT INCENTIVES

It is important to provide a reasonable (but not coercive) incentive to respect the time that your participants spend taking part in your study. Over the years, we have experimented with a wide variety of incentive structures for reimbursing people for their participation in our studies, and we have found that what works best can vary depending on who you need to recruit.[4] Your institution likely has rules about the amount of incentive you can offer, how you can offer it (e.g., gift card, cash, food, technology, etc.), and to whom you can offer it (e.g., you may have restrictions on who can participate, or at least who can receive an incentive), so always check with the experts at your institution before advertising or finalizing your incentive plan for a study. With that in mind, you should ensure that you are paying participants a reasonable amount for participating, and consider travel time and other costs they might incur because they participated in your study (e.g., will they incur an additional data charge with their smartphone's service plan?). For example, a one-hour interview may take three hours of a person's time if they have to travel to your lab or office from far away or via public transportation. General industry practices at the time of printing generally pay participants $US 100 for a one-hour session in a lab or $US 200–250 for a multi-week field study with two interviews. It is not unusual for professional recruiters to turn away your job if they don't believe you're paying a fair incentive to participants for what you're asking of them. While academic incentives are often lower than this, we think it best to use the reference of industry standards.

---

[3] The cost can be higher in some situations, particularly if you're looking for hard-to-find or hard-to-schedule participants (e.g., small business owners).

[4] For example, in a study where we needed to recruit high-paid executives, they were more interested in us making a charitable donation in their name and receiving a neat piece of swag that couldn't be purchased anywhere else than they were with a monetary incentive.

There are a few lessons that we have learned in pricing longer diary and field studies (we discuss one example in detail in Chapter 4). First, make sure that your incentive isn't tied to what you're trying to study. For example, if you want to study how someone will use your system, their incentive shouldn't be based on using your system. If you were using a diary study (see Chapter 4) to learn about system use, they should be paid for diary entries that report no use of the system (in fact, you should encourage that, as they'll likely want to use your system just to be helpful). If you want to see if your system encourages people to perform a behavior like meditating, their incentive shouldn't be tied to meditating.

Second, it is often a good idea to incent participants on a sliding scale based on their level of participation. This can take some experimentation, as paying explicitly per-entry can lead to participants skipping some entries as each might only be "worth" a few dollars (as explored in detail in Ariely, 2008; we also discuss this a little more in Chapter 4). That said, it is also not advisable to make incentive schemes all-or-nothing, or with a large cliff after missing one or two entries. This can disincentivize participants to continue if they have missed a small number of entries (which is bound to happen with people's busy lives). As already mentioned above, it is important to clear whatever incentive plan you want to use with the experts at your institution to ensure you are complying with your institution's rules.

1.1.5   FIELD STUDY LOGISTICS

If you are going into the field to meet with or observe a participant in their home, it is important to make sure that everyone (i.e., the participant(s) *and* the researcher(s)) feels comfortable. It is often best to attend a home visit as a pair of researchers, both for the researchers' and participant's safety and ability to feel at ease. Larger research teams might be intimidating, and lone researchers can raise feelings of discomfort, especially if of the opposite gender of the participant. We also find that it's often best if the home visit team consists of one male and one female researcher, again so participants (and the researchers) feel more comfortable. It can be helpful to ask participants in advance about any animals in the home, especially if one of the researchers has allergies or fears. You might also ask participants to make sure loud dogs are put outside or muzzled to help keep audio recordings clear and to help researchers feel safer while visiting the home. You should be prepared to follow their household rules (e.g., you might have to take your shoes off at the entry). We usually make sure someone on the research team who isn't coming on the home visit knows where we are and when we should be finished (again, for safety reasons).

You'll probably also want to bring extras of any tools that you're using to collect data (e.g., we like to have extra audio recorders, batteries, notepads, pens, etc. on hand in case we run out of anything or anything breaks).

If you are going into the field to meet with or observe a participant outside of their home, make sure you already have whatever permissions are needed to conduct your research (e.g., if you'll be in a store as part of your research, you may need to clear that with the store's management prior to showing up). Contact the experts at your institution well in advance of your study to find out about what permissions you'll need and how to obtain them.

1.1.6   ETHICS, CONSENT, AND REVIEW

Regardless of the specific policies of your institution, it is important for both you and your participants to have your study plans reviewed by experts, including ethics experts. Even the most experienced researchers may not consider some aspect of data collection, storage, results, or privacy; having experts review your detailed study plan, as well as the write-up of your results, can help to identify these issues and ensure that you conduct and report on your research appropriately. In many locations, there are a variety of legal concerns and best practices around obtaining consent that must be followed.

Most institutions have some form of a review board, often called an *Institutional Review Board* or *IRB* that reviews study proposals.[5] Generally, you will need to prepare a document about all study procedures, including a list of questions that will be asked, and information about how data will be stored and analyzed (including for how long you intend to retain the raw data, and how you plan to dispose of it when the retention period is over). If you are running a field study of a system, details about how the system stores and transmits data, and the types and quantities of data it collects will also be a part of this document. A panel of senior researchers will typically meet at least monthly to review study proposals, and may ask for clarifying information. It is critical to prepare plenty of time for this review process as it may take several rounds to obtain approval, and you may not even begin the process of reaching out to recruit potential participants until the proposal is approved.

Part of this process is preparing a consent form for participants to sign prior to starting your study. Consent forms are important so that participants understand their obligations for participating, how they will be compensated for their time, any data that might be collected, the risks and benefits of participating, and the circumstances under which you may use the data you collect (e.g., whether you may show photos taken of the participant in presentations of your research in academic/professional venues, whether those materials can be posted online, etc.). Often, your institution will have a template that you can start from that includes sections that have been pre-approved by your IRB or legal team. For some simple usability studies, there may be forms that can be reused

---

[5]   If you don't have something called an IRB, it doesn't mean that you're exempt from review, and even if you are exempt from review, you must still follow ethical and legal research procedures. Check with others at your institution to ensure you know who from your institution should review your study plans (it may involve several departments—e.g., legal, public relations, etc.), and what rules and procedures you need to follow.

in their entirety. There are several helpful resources about obtaining informed consent (e.g., the APA's *Ethical Principles of Psychologists and Code of Conduct > Standard 8: Research and Publication*[6] and the U.S. Department of Health and Human Services' *Informed Consent Tips*[7]). Always consult your institution on the practices to which you should adhere.

In general, it is best to consult with as many experts as possible. If you are transmitting and/or storing data, consult security and legal experts. If you are conducting interviews or surveys, consult experts on these methods for best practices and information on how to best remove personally identifiable information, anonymize data, and store it on a device/service that is approved by your institution. If you are storing medical data or working with underage participants, regulations such as HIPPA[8] or COPPA[9] might apply to how you collect and store data. Research that involves collecting sensor data might benefit from information from The Core.[10] In addition to legal obligations, it's also important to consider your ethical obligations to your participants. If you are interviewing multiple members of a family or a group of friends, the data that one person tells you should not be shared with other participants. If you are in a home and another family member or visiting friend wants to participate in the interview, make sure you have a plan for handling that situation (e.g., by having an extra consent form for them to complete before recording or interviewing them, or by politely declining their offer). As an example, in the "Method" section of Matthews et al., (2017), we discuss many of the ethics-related steps we took in our study of the privacy and security practices of survivors of intimate partner abuse.

We cannot stress enough the importance of following these procedures. Your participants are trusting you with their information (and possibly the information of others), and you need to ensure that you are doing all you can to keep that data safe and secure. On top of this, there are legal obligations with serious consequences in many locations, and rules often differ across jurisdictions—e.g., what's acceptable in one country may not be in another. Don't go it alone. Consult the experts.

1.1.7   PILOT AND REVIEW EVERYTHING

While this basic introduction to user research provides you with a foundation for choosing the right methods, asking the right questions, and finding reliable conclusions from your data, we again stress the importance of consulting experts and reviewing all aspects of your study (including piloting all aspects of your study—a point we return to later). Beyond the basics of following required review

---

[6]   http://www.apa.org/ethics/code/index.aspx?item=11 {link verified Dec 26, 2016}.
[7]   https://www.hhs.gov/ohrp/regulations-and-policy/guidance/informed-consent-tips/index.html {link verified Dec 26, 2016}
[8]   https://www.hhs.gov/hipaa/ {link verified Dec 26, 2016}.
[9]   https://www.ftc.gov/enforcement/rules/rulemaking-regulatory-reform-proceedings/childrens-online-privacy-protection-rule {link verified Dec 26, 2016}.
[10]   http://thecore.ucsd.edu/ {link verified Dec 26, 2016}.

processes, ensure that you have consulted with the appropriate legal, security, research methods, and technology experts to ensure that participant data is secure and stored appropriately and that you are conducting the research and analyzing the data in a way that will yield valid, reliable, trustworthy results.

Now let's get started!

CHAPTER 2

Sensor and Usage Data

Mobile devices are exciting platforms for creating new types of experiences because of the rich combination of sensors and hardware that are available on the device. From sensing location, motion, proximity indoors or to other people, activity, or just what users are doing in an application, mobile devices enable an entirely new range of interaction and research opportunities. This chapter will discuss sensors, some interaction and research opportunities they afford, and how you can use these data streams to better understand participants and users.

On-device sensing, as well as the increased prevalence of ubiquitous sensors (e.g., wearable technologies such as smartwatches or internet-connected devices such as smart thermostats), are changing the ways that we think about well-being and environmental control and are allowing billions of people to unobtrusively sense daily activity such as step counts, or providing near continuous logging of heart rate. Other technologies such as Bluetooth Low Energy (BLE) beacons are enabling accurate indoor positioning and interesting opportunities for retail and museum education (Pierdicca et al., 2015; Martella et al., 2016). Together, these various sensors are creating new opportunities for studying human behavior "in the wild," facilitating behavior change via digital technologies, and creating "citizen science" activities, whereby individuals are actively engaged in supporting the documentation and understanding of both man-made and naturalistic ecosystems.

In addition to sensing the world, mobile technologies can enable researchers to record user's behaviors and to understand how new services fit into their lives. We can explore when and where applications are used as well as the specific paths that people take throughout the screens of our applications. We can discover places in the design where users get stuck or abandon the service as well as run randomized experiments on large numbers of users to understand the impact of specific design changes. While instrumentation and experimentation offer many possibilities for design and product development, they can also raise a host of ethical issues depending on what is being manipulated.

The goal of this chapter is to enable you to understand a wide variety of sensors and logging tools that are available when you're crafting a new mobile experience and seeking to understand use in real world situations. We will provide a summary of different sensor data sources, particularly smartphone sensors, digital traces, contextually embedded sensors, "wearable" technologies (e.g., fitness tracking devices or smartwatches), and other emerging "Internet of Things" (or "IoT"). As the mobile and ubiquitous sensing landscape is changing rapidly, this review cannot be exhaustive, but instead the goal is to provide you with enough information to support thoughtful selection of sensors for your mobile user research endeavor.

The chapter provides an overview of various data types and then discusses a variety of use-cases for data from these sensors, factors to be mindful of when selecting sensors, and practical suggestions for balancing the tradeoffs that are often prevalent when using these sensors. The chapter closes with a variety of practical guidelines to consider when using mobile sensing or instrumentation, as well as a variety of limitations of these technologies.

2.1   DATA TYPES

Overview: There are a variety of data sources that can be used for mobile user research. Table 2.1 provides an overview. While there is overlap in these data sources, broad categories include data from smartphone sensors, digital traces, and contextually embedded digital sensors (e.g., wearable technologies and the emerging IoT). Each of these data sources will be discussed in turn. For all of these data sources, it is important to be clear about the difference between the *raw data signals* that are measured compared to the *concepts that can be inferred* from these raw signals. For example, a 3-axis accelerometer is a sensor that detects movement and acceleration in the form of raw "G's" or the degree of pull that occurs from both acceleration and gravity across three different axes to enable understanding of movement in 3-dimensional spaces. This raw signal has been used to infer a wide range of concepts such as "steps" (which are inferred by measuring the peaks that are observed within raw accelerometry data when a person takes a step), sleep (inferred by a pattern of overall lack of movement while a person wears a wrist-worn accelerometer), and even more complex behavior such as cycling, which has a relatively unique raw accelerometry profile (Bao and Intille, 2004).

The distinction between what is measured vs. inferred is important because, for many of these sensors, few standards exist, thus making comparison across seemingly similar measures often very difficult. For example, the auto industry has a long tradition of standardized measurement of concepts such as speed, fuel consumption (e.g., miles per gallon), or acceleration. To a lesser extent, good standardization of measurement is also available for some medical devices, such as measurement of blood pressure. That said, the majority of measurement targets for mobile user research are still in their infancy, and thus few standards are currently available (e.g., different ways of inferring "stress" in context). For example, there is no clear standard on what exactly constitutes a recorded step of a person via a wearable device. This results in different estimates of steps from one wearable technology to the next (Floegel et al., 2016). These variations in estimates exist because of different hardware specifications (e.g., different accelerometers with different sensitivities), different algorithms used to translate raw signals into inferred concepts (e.g., different thresholds for defining when a "peak" can be inferred to be a step) and even differences in manufacturing of the same hardware (Floegel et al., 2016). Based on the distinction between what is measured vs. inferred, we first describe some sensors that are currently available, and then we discuss some of the common concepts that are currently inferred from those sensors. In the second section, we also discuss what can be learned by instrumenting interaction on the device itself and how analytics data can be used to improve the design of mobile systems.

Table 2.1: An overview of common sensors available in current mobile and ubiquitous computing devices

Sensor	Type	Usage examples
Motion (e.g., gyroscope, accelerometer, compass)	Smartphone Sensor	Motion sensors can detect activity (walking, driving, biking, standing) as well as infer larger activity states such as daily step counts.
Location (e.g., GPS, cell tower ID, Wi-Fi fingerprinting, Bluetooth beacons)	Smartphone Sensor	Location sensors can frequently locate users into a particular building with high accuracy. Beacons can place users even more precisely within an indoor environment.
Environment (e.g., temperature, humidity, barometric pressure)	Smartphone Sensor	Many mobile devices come equipped with environmental sensors. These can be used to detect when people go in/outdoors or correlated with other sensed data.
Microphone	Smartphone Sensor	The microphone can be used for detecting anything from speech to moments of social activity, television watching, or eating/chewing.
Camera	Smartphone Sensor	Cameras can be used to capture photos and videos, but also to check on ambient light levels, to perform object recognition on items in a scene, or to scan QR codes.
Social Interactions (e.g., phone call or SMS logs)	Digital Traces	Call and SMS logs can be used to infer overall social health as well as to better understand the relationships in a user's life.
App usage (e.g., launches, installs, durations of use)	Digital Traces	App usage can provide researchers access to start and stop times of use of any app installed on a device, which can be useful for understanding other apps that co-occur in sessions with your app.
Web history	Digital Traces	Web history can provide a list of all URLs visited on the mobile device.

Sensor	Type	Usage examples
Media sharing (e.g., messaging, social media)	Digital Traces	Traces of media sharing can identify topics that are important to the user as well as aspects of the social networks with which they share.
Home context (e.g., heating controls, light use, door locks)	Contextually Embedded Sensor	Smart home sensors can provide a wealth of information about one's context and potentially activity (e.g., watching television, cooking).
Food Consumption (e.g., smart refrigerators, coffee pots, trash cans, toilets)	Contextually Embedded Sensor	New in-home sensors allow increased understanding of food-consumption behaviors such as when and what a person eats.
Heart rate	Wearable Sensor	Most modern smartwatches enable continuous heart rate monitoring throughout the day and during workouts.
Mood/arousal (e.g., electrodermal activity)	Wearable Sensor	Electrodermal activity can measure the intensity of one's arousal over time, which is often correlated with specific moods.
Arm motion/steps (e.g., wrist-worn accelerometers)	Wearable Sensor	Many wrist-worn devices enable capturing motion of the device, which can be used to infer activities such as walking, step count, or sleep.
Respiration	Wearable Sensor	A variety of wearable sensors can measure breathing activities.
Blood glucose sensors	Wearable Sensor	New implanted devices, patches, and other devices are increasingly able to detect a person's current blood glucose levels, critical for those with diabetes.

Smartphone Sensors: The modern smartphone includes a truly impressive array of different sensors that can be used for mobile user research such as accelerometers, gyroscopes, global position system (GPS) receivers, Bluetooth Low Energy signals (BLE), digital compasses, temperature gauges, microphones, and high-resolution cameras. Beyond the sensors, valuable data are also available related to social interactions via smartphone's text messaging and telephone capabilities as well as timestamps from clocks within smartphones that are increasingly synchronized to a standard time metric, enabling comparability of experience across data sources for a single person over time and across individuals. Finally, a variety of usage data, both of the device and of specific apps, is also available for enabling further study of interactions in the wild.

These sensors and usage logs in smartphones are increasingly being used to infer a wide range of human behaviors and other concepts relevant to mobile user research. For example, there is increased sophistication in tracking various types of physical activity, particularly steps via accelerometry, speed, and distance traveled while walking, cycling, or driving/taking mass transit via GPS and even triangulation via cell tower signals (Fan et al., 2015).

Other emerging inferred concepts include:

- estimates of emotion, stress, mood, and even psychopathology such as depressive states (Canzian and Musolesi, 2015);

- social interactions via the use of microphones to infer speech in context or via communication logs, such as SMS;

- co-location of individuals within a given context via bluetooth and/or GPS (Madan et al., 2010; Wyatt et al., 2011; Lu et al., 2009);

- eating habits via the use of microphones to infer chewing (Amft and Tröster, 2009) or via processing of photos of food (Martin et al., 2012; Harray et al., 2015);

- more specific activity types (such as driving or sleeping); and

- commuting and transit patterns via GPS (Harari et al., 2015).

Beyond these factors, diagnostic tests are increasingly being explored via the use of the smartphone. Examples of this type of data stream are those being developed as part of the Apple ResearchKit. From signals already discussed like physical activity (MyHeart Counts) to inferring tremors among Patients with Parkinson's (mPower) and other cardiovascular health indicators such as heart rate or blood oxygen saturation (via the camera), the list of plausible targeted concepts being inferred is truly impressive.

When using this type of data, and particularly when training machine-learned models, it is critical to gather representative training sets, such that the models are representative of a wide range of uses. When developing the Mobile Sensing Platform at Intel and the University of Washington, we discovered that our initial set of (mostly male, mostly younger) participants did not walk in the same ways that many of our later participants did. In the end, we had to retrain a series of sub-classifiers for activities such as walking in high heels or walking in a variety of different environments such as in a store vs. on a fitness trail (Choudhury et al. 2008).

Digital Traces: There is also movement toward inferring psychological, social, and contextual variables from "digital traces," meaning the information gathered while using digital tools such as email, social media, web browsing history, movie viewing, etc. These data are either logged interactions (e.g., downloading of applications, purchasing history) or freeform text that are sometimes tagged (e.g., #YOLO). Logging of mobile interactions within your app, otherwise called app instrumentation, is a unique example of digital traces that can be so valuable to mobile user research that we will discuss them separately at the end of this section. In particular, these logging data can be evaluated with concrete metrics that can be optimized for improving flow and usage of your mobile system.

Interaction data, particularly related to search and shopping behaviors, have been used to infer a wide range of concepts, particularly various user preferences that translate into recommendations (e.g., shopping suggestions within Amazon or movie recommendations with Netflix; Resnick and Varian, 1997, or a movie recommendation based on your previous movie watching history). In terms of the more free-form text, interesting information can often be gleaned from the various tags themselves (Brownstein et al., 2009); there is also a wealth of interesting information that can come from natural language processing of freeform text. Natural language processing (NLP) is a sub-field of computer science focused on better understanding the interaction between computers and natural human languages. The field has a wide range of techniques available for processing free text (both written text and audio) into meaningful data for a wide range of purposes. This includes highly advanced natural language processing such as Apple's Siri or IBM's Watson, but there are also several types of NLP that focus more explicitly on parsing free text into plausibly meaningful data points. For example, a classic tool of very basic NLP is the Language Inquiry and Word Count (LIWC, pronounced "Luke") program. In this program, dictionaries are created that monitor various types of words, such as nouns or verbs, or words that are related to topic areas; for example, "run" is a word that could plausibly be related to health. With these dictionaries, the system then counts the number of instances of terms in each dictionary.

NLP can also be used to parse out phrases, sentiments (e.g., positive or negative tone), or syntax-related issues within freeform text to name but a few. Sentiment analysis has been used, for example, to detect the mood of the Internet on particular topics via Tweets in Twitter Weather (Holtzman and Kestner, 2017). In this system, topics were displayed as a "weather forecast" such that topics with positive tweets displayed as nice warm and sunny days, while topics with negative sentiment were cold and snowy. Methods such as TF-IDF (term frequency–inverse document frequency) can also be used to detect rare words in particular corpi, such as users who talk more about specific topics than the general public.

These various ways of coding freeform text enable a wide range of concepts to be inferred such as social status and hierarchy to personality characteristics and mood (Adalı and Goldbeck 2014; Estrin, 2014; Golbeck et al., 2011; Hekler et al., 2013b; Pentland, 2014; Tausczik and Pennebaker, 2010). The tools can also be used for social networks and interactivity between individuals (Lazer et al., 2009). When combined, particularly with other streams of data, these data provide an interesting assessment of the daily life of individuals that can provide more rich understanding of mobile interactions or for studying behavior itself.

Contextually Embedded Sensors (The IoT): There is a growing movement in the IoT, which involves the digitization, often via embedded sensors that are linked to the Internet, of everyday appliances and devices used in our homes (e.g., "smart" refrigerators), work (e.g., "smart" white boards), and commuting environments (e.g., internet-connected cars). The target of this work is to "create 'a better world for human beings', where objects around us know what we like, what we want, and what we need, and act accordingly without explicit instructions." (Perera et al., 2014). From "smart" thermostats (e.g., Nest), refrigerators (e.g., Samsung's refrigerator that includes cameras within the fridge to allow one to check what food is in the fridge from a smartphone), and cars, the technologies we interact with everyday are increasingly connected to the Internet, thus supporting an enormous explosion of data sources. In terms of the actual sensors involved, the breadth of different IoT devices is enormous and thus we do not include a list of actual sensors for such a heterogeneous class of devices. That said, as with smartphones and digital traces, we urge you to be mindful of the difference between what is measured vs. inferred, even in these devices. Within the realm of IoT, however, many sensors already have robust standards (e.g., speed of cars) and thus the core advancement here is the ability to access these data streams. This connectivity of data across various data sources enables the sort of machine learning or pattern recognition explorations increasingly being examined under the moniker of "data science." For several of these sensor classes, the main innovation is in connecting robust and standardized sensors to the Internet to support integration of these data sources for more complex insights.

In terms of what is inferred from these sensors, the types of information being gleaned is truly staggering. For example, all major automobiles include a plug-in source for downloading a wide range of data related to driving patterns including speed, distance traveled, issues with the engine, and other factors. Some insurance companies take advantage of these data to generate driving profiles that can be used to develop more robust predictions on the likelihood of traffic accidents and, by extension, cost estimates per driver. Further, unique technologies are being developed related to health. For example, there is increased interest in exploring the incorporation of sensor technologies into toilets to automatically measure biomarkers and plausibly the microbiome profiles of individuals (Turgeman et al., 2014), or the use of facial recognition features within bathroom mirrors to assess health problems or the monitoring of alcohol or tobacco use (Colantonio et al., 2015). "Social sensing" is also a particularly interesting topic area for further work (Mast et al., 2015). Within this domain, the focus is on sensing and inferring a wide range of interpersonal and behavioral interactions between individuals such as head nods, eye gaze (as a metric of focus and attention), gesticulations, eye contact, and emotional inference. These inferences are often made from a variety of sensors, in particular, audio and/or video recordings, the Kinect sensor (which can infer body movement in more detail), and increasingly via the microphones embedded in smartphones for more ecologically valid extraction of insights.

One other sensor class that is particularly relevant for mobile user research are beacons that can be used to detect proximity to locations or objects that are tagged with these beacons. While different technologies are available for this (e.g., BLE, RFID's) both iOS and Android have incorporated protocols for detection of BLE beacons with increased commercial technologies enabling this (e.g., Estimote). While the data type is simple (e.g., timestamp of being in proximity to a given beacon or signal strength to each beacon, allowing for triangulation of location with multiple beacons), these beacons are increasingly small, which can enable their placement within doors but also on other people or objects (e.g., pets) thus enabling a rich understanding of a person's activities.

Wearable Technology: A unique type of contextually embedded sensors is "wearable technologies." Wearable technologies are clothing and accessories (e.g., wristbands, watches, necklaces, head or chest bands) that incorporate digital tools such as microprocessors, data transmission features (e.g., Bluetooth-connected devices), and sensors (e.g., accelerometers) to achieve some specific need (e.g., easier interactions that do not require phone via a smartwatch, supporting fitness goals via activity trackers like the Fitbit). Perhaps the most popular type of wearable technologies today are wrist-worn fitness trackers such as the Fitbit Surge, Microsoft Band, and smartwatches. These devices include an impressive array of sensors. For example, the Microsoft band includes: optical sensors often used for inferring heart rate, accelerometers, gyrometers, GPS, ambient light sensors, skin temperature sensors, UV sensors, touch screens, galvanic skin response, microphones, and barometers. There are also other devices that are designed specifically for more medical purposes such as "patches" that use electrodes and concepts from electro-chemistry to infer estimates of glucose in a person's blood (McGarraugh, 2009).

Similar to data from the smartphones or digital traces, these raw signals are increasingly being used to infer a wide range of concepts. One of best-developed areas for wearables is in tracking physical activity and sleep. As evidenced by the use of the patch, there are a variety of medically valuable targets increasingly being inferred such as continuous glucose estimates, HbA1c (an estimate of overall glucose content in blood over three months), and a wide range of data about cardiorespiratory fitness such as pulse, heart rate variability, respiration rates, and blood oxygen saturation. These wearable sensors thus enable highly personal information to be gathered about a person, often in near real-time.

2.2　GENERAL USES OF THESE DATA

Sensor data and the concepts inferred from it offer exciting opportunities for mobile user research. Here we briefly provide an overview of five broad (and again overlapping) areas of use: the study of naturalistic behavior, facilitation of behavior change, natural environments and complex systems, improving mobile experiences via app usage data, and understanding overall device use.

Many of these uses rely on exploring large amounts of personal data for individual users. Popularized by Deborah Estrin, the concept of "Small Data" (Estrin, 2014) is gaining traction in the research community. In comparison to "Big Data" which often analyzes interactions from millions of users, *small data* looks at the data of an individual over time to find patterns. This can include finding opportune times to suggest physical activity, automatically learning important locations in people's lives from data traces (Choudhury et al., 2008), or finding patterns across a wide variety of daily activities (such as gaining weight on rainy days) (Bentley et al., 2013b).

Many people now have a large amount of data that is collected about their lives. However, this data is rarely used to its full potential. All owners of an Apple Watch, for example, have up to the minute step counts as well as frequent heart rate readings that mostly go underutilized. And location traces of the places that people visit are rarely utilized, and almost never combined with other types of data.

The concept of small data asks us to move beyond data silos and examine all of the data that is collected about our individual lives. This is in contrast to so-called "big data"—or the aggregated usage data collected from millions or billions of users of a deployed system, such as Facebook, Netflix, or an email provider. The sections below highlight uses of both "small" and "big" data for understanding how people engage with technological systems and the outcomes that they experience after doing so.

Quantifying Use of Your System: The most common use of logged data in mobile user research is in understanding precisely how people are using your system in the real world, outside of artificial conditions of a lab environment. Usage logs can often become essential metrics for understanding the usefulness of your system and can provide a means for testing the impacts of design changes over time on key metrics such as Daily Active Users, Session Length, etc. If you are conducting an evaluative field study (discussed in Chapter 3), you may want to include some degree of instrumentation so that you can learn how a system is being used in the wild. There are a variety of ways to capture data and many important considerations to ensure that a participant's privacy and security are being protected while collecting the usage data that is required to improve the service or understand its use.

In general, there are two types of interactions that are most frequently logged in mobile applications. Most commonly, researchers may want to capture each screen view in order to understand a user's flow through the application and which areas of the system are used the most, or used in specific contexts or times of day. Secondarily, particular actions or events within the application can also be logged. These can include within-screen interactions such as scrolling, enabling or disabling particular settings, or clicking certain buttons. System-generated events, such as refreshing data sources, receiving a push notification, or running a particular background operation can also be logged in this way. A combination of screen-based logging and action or event logging can provide a deep picture of how participants in your study are interacting with your system and will help you to redesign your interactions to modify behaviors in the wild.

There are many ways to capture the raw data that you will need. In general, you will want to store some unique user or device identifier, a timestamp, an identifier of the screen being viewed, and the action that was performed. It's quite easy to implement this yourself, cache events, periodically send them to a server, and log them into a text file or database that can be analyzed. There are also other tools available to make logging easier, such as Flurry (from Yahoo) or Google Analytics, which enable you to log screen views as well as events and provide online dashboards to follow and understand use. Both of these platforms are designed to be up and running in a matter of minutes, with no more than an hour needed to fully instrument a basic application.

No matter which way you choose to get the data to a server, it's important to consider privacy and security concerns and other important factors about the data that you are collecting. If you are collecting the location of your participants, or other sensitive data, you may want to perform steps such as upleveling location (e.g., to neighborhood-level data) so that you cannot identify residences or places of work. If you are taking screen shots, be sure that you don't capture participants' authentication credentials. If you are logging communications-related data, remember that there are two parties to a communication and some states, countries, or Institutional Review Boards require two-party consent to capture certain details of communications. Also, when logging URLs, often private or identifying data can be included in the address, such as a user name on a particular service or hashcode to access a "private" file in Dropbox. In commercially deployed systems, you probably also want to ensure that your user or device IDs cannot be associated back to a particular user. Obviously, for small-scale studies where you meet each participant in person, you will want to be able to associate these logs back to specific individuals so that you can interview your participants about the specifics of their experiences. Consider every piece of data that you plan to log, and what types of information can be present in that data, and ensure that you have the participant's privacy and security in mind when deciding what to capture and store.

Your IRB or legal team may have specific recommendations on how to convey what is being logged to your users or research participants. We often show potential participants in our research studies examples of the types of data that we will collect when they review the consent form to participate in one of our studies. For some types of studies, you may want to have a mechanism for participants to view and edit their data before it is sent to the research team (e.g., if they want to remove links to certain sites from a web history, or not share a particularly personal conversation or photo with their partner in chat logs). Understand that your system will be a part of their lives, and that most people have aspects of their lives that they do not want to share with a research team. Often, IRBs and corporations have specific rules about data retention, and you should be careful to abide by these as well.

Once you have collected the data, there are many types of analysis that you might want to perform. One of the most basic is to look at a "funnel" of use. A funnel starts with every instance of opening your application and then shows which percentage of users traveled to particular screens, and in what order. It is called a "funnel" since with each level of screen transition, you will invariably lose users who exit your application to do something else. Analyzing a funnel can show you where you lose the most users, and in these cases there is often a great opportunity for simple changes (e.g., wording of a prompt, changing of an icon, etc.) to improve engagement—assuming that's the goal. Once you discover one of these points of exodus in your application, often conducting an in-lab usability study (see Chapter 3) can help you identify precisely why people are leaving at that point. These venues can also help you test new concepts and see if participants understand the desired flow more accurately.

Analytics data also allows you to study "retention," or the number of users who return each day to your application, or who use a particular feature repeatedly. There are a variety of ways to measure this. Most simply, you can look at daily retention for users who first came to your app on a particular day. How many returned the next day? And the day after that? Frequently, 7-day retention numbers are used as a good metric for engagement. When looking at these numbers for your own application, do not be discouraged. Data has shown that 23% of users will only open an application once (O'Connell, 2016), and that 30-day retention rates are well below 10% for some categories of applications (Klotzbach, 2016).

Another way to look at retention is on a per-feature basis. For our *Health Mashups* study (Bentley et al., 2013b), we plotted each user as a column and looked at their use of each feature over the 90 days of our study. Figure 2.1 shows the results. It is important to note that in this study, participants were not asked to use the system; rather, they were explicitly told that they should use it as if they just downloaded it from the app store. No compensation in the study was tied to use. We have seen this work well in creating natural use patterns and, in other studies of less-successful systems, we have seen use drop to zero almost immediately.

The goal of the Health Mashups system was to present data via natural language, so that users did not have to engage with complex graphs, and we can see that this worked, with the majority of interactions going toward logging or viewing the natural language observations. We can also see several participants who abandoned the app, and we can see which actions they performed before doing so, perhaps giving us clues about what to change in future iterations to keep these people engaged before enough data is collected for statistically significant observations to be available.

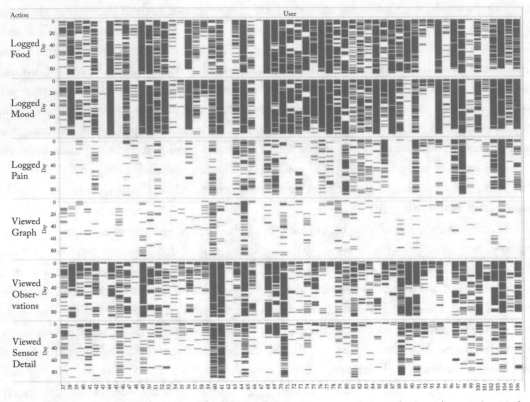

Figure 2.1: The usage of each feature of the Health Mashups system by user (column) over the 90 days of the study. Several different usage types emerge when looking at this figure, which could then be combined with qualitative data from interviews and diary entries. From Bentley et al. (2013b).

2.2.1   UNDERSTANDING OVERALL DEVICE USE

Beyond understanding interactions with your particular application, sensor and usage data can help you understand use of the entire mobile device. This can be quite important when trying to characterize the amount of time that participants spend in specific applications or specific categories of applications, or when and where they use these applications in their everyday lives. This can be important for strategic, generative field research when trying to understand if an idea you have fits a real user need (see Chapter 3 for more on generative field research).

Böhmer et al. (2011) pioneered the use of app logging to understand device use over time. As the title of their paper suggests, through analyzing app use by time of day, they saw participants literally "Falling Asleep with Angry Birds, Facebook, and Kindle." Their Android application included a component that would receive system-wide *BroadcastIntents* when any application was opened or closed. They were able to analyze categories of apps used, average times of use for specific apps and categories, as well as times of day of use.

We deployed a similar app logger in our study of teenagers' use of mobile phones (Bentley et al, 2015). For this study, both an Android and iOS app logger were deployed, and we were able to understand how and when teens were using an average of 9.5 different communication applications in their lives. We combined these logs with qualitative interviews and voicemail diaries to more broadly understand why participants were using all of these applications and how each fit into their lives in a unique way as well as opportunities that existed for new communications applications.

Pielot et al. (2014) developed a different type of app logger that kept track of Notifications on the device. Through this, they were able to understand the types of notifications that different apps create, when people receive notifications, and how they respond to them. This has important implications for how new applications should notify users, a timely topic for mobile application design.

In other examples of understanding mobile device use, we have captured a participant's mobile web-browsing history or asked them to provide examples of link sharing in mobile messaging (Bentley et al., 2016). These studies have helped us shape new product strategy by understanding how people use the mobile web and how they share links with others. New concepts around link sharing and messaging were created as a direct result of this data, as we discuss in the paper.

System-wide application or notification logging can help you understand how your system fits into the wider ecosystem of apps on a participant's device. You should be careful to make sure that your participants know that you are doing this as a part of the informed consent process (see Chapter 1 for more on informed consent). We typically show participants examples of exactly what log data we will collect (see Figure 2.2 for an example).

P1,Mon Jul 08 12:23:12 PDT 2013,Chrome,149
P1,Mon Jul 08 12:25:40 PDT 2013,display off,2156
P1,Mon Jul 08 13:01:36 PDT 2013,Chrome,2
P1,Mon Jul 08 13:01:37 PDT 2013,display off,2
P1,Mon Jul 08 13:01:39 PDT 2013,Chrome,7
P1,Mon Jul 08 13:01:45 PDT 2013,Email,5
P1,Mon Jul 08 13:01:49 PDT 2013,display off,2593

Figure 2.2: Example logs from a system-wide app logger shown to participants as part of the consent form for participating in the study. Each line contains the participant ID, timestamp of application start, name of the application, and duration of use.

Naturalistic Behaviors and Outcomes: The sorts of questions that were classically explored within the field of epidemiology such as estimations of incidence and prevalence of topics of interest (e.g., infection rates) as well as trends in these topics over time are now being explored using mobile devices, but with added temporal and contextual elements that these new data sources afford. As with diary studies and experience sampling (see Chapter 4), the questions explored can often be organized by the relevant topic area (e.g., health, psychology, or sustainability).

Smartphone sensors are increasingly enabling research questions and topics of factors that are sensed and inferred purely from the phone. These include explorations into the relationship between social interactions and worksite interactions (Mast et al., 2015; Pentland, 2014) or the movement of individuals through urban landscapes (Griswold et al., 2004; Harari et al., 2015). There are also interesting opportunities for melding experience sampling techniques with sensors, particularly with the use of sensors as trigger points for context-sensitivity, such as the use of geo-fences of GPS data that define specific locations as "home," "work," and so on as triggers for launching questionnaires (Dunton et al., 2016). The movement toward use of Apple's ResearchKit also enables the naturalistic study of behavior and the relationship to important outcomes such as health (Bot et al., 2016).

With regard to digital traces, the Computational Epidemiology Lab at Boston Children's Hospital has conducted landmark work in this domain that is making full use of various digital traces, particularly those available from Google Search. For example, the group has conducted research to analyze patterns of influenza epidemics, understanding geographic patterns of substance abuse, and examined the impact of pollution on chronic disease symptoms (Brownstein et al., 2010; Santillana et al., 2015). More recently, they have explored the use of a smartphone and related sensors as a tool for public health surveillance related to health outbreaks (Herrera et al., 2016).

In terms of contextually embedded sensors, most tools are still largely in their infancy and thus there are few exemplars. That said, there are some more classic epidemiology studies that might point to plausible research questions to explore with embedded sensors. For example, the National Health and Nutrition Examination Survey (NHANES), which is a nationally representative survey focused on better understanding the relationship between behavioral and psychosocial phenomena on health outcomes, has increasingly incorporated the use of the ActiGraph accelerometer for monitoring physical activity. The breadth of questions that are explored in NHANES is a logical exemplar of the sorts of questions that could likely be explored with the increased availability of contextually embedded sensors, particularly wearables, but also valuable data for understanding eating behavior in context. As a second example, in August 2014 there was an earthquake in Northern California. Jawbone utilized the data that monitors the sleep activities of their users to examine the impact of the earthquake based on geographic region as measured by percent of users who were awakened by the Earthquake (Mandel, 2014). Overall, we are still just starting to understand the possibilities for studying naturalistic behavior and outcomes.

Behavior Change: There has been rapid growth and interest in the use of these technologies for fostering behavior change. Among the early examples of the use of augmented sensors for fostering behavior change is our UbiFit (and later UbiGreen) system (Consolvo et al., 2008a, 2008b;

Froehlich et al., 2009). These systems included separate sensor arrays to detect physical activity patterns and utilized this information to inform the design of an evolving "glanceable display" wallpaper on a phone (described in Chapter 4). This basic concept of translating passively collected physical activity data into visually appealing background and related feedback has now been replicated in a variety of other ways such as the MILES study (described in Chapter 6), which utilized passive tracking of physical activity via the phone's internal accelerometers (King et al., 2016) to provide feedback either in the form of social feedback (e.g., activity relative to others), affective feedback (i.e., feedback in the form of a bird avatar that flew faster, was happier, and sang songs as you walked more), or more of a cognitive feedback (i.e., established goals and then received feedback on level of activity achieved relative to those goals).

Another example of a smartphone-based behavior change technology is our Health Mashups system (Bentley et al., 2013b). In this system, a variety of data, most of which was either inferred via the phone or some other wearable/nearable technology (i.e., weight, sleep, step count, calendar data, location, weather, pain, food intake, and mood) was collected and then basic correlations between the data were calculated. Insights from those correlations were shared with participants in natural language (e.g., "You are happier on days when you sleep more."). An even more recent example is the MyBehavior System (Rabbi et al., 2015). This system utilized data from the smartphone to understand naturalistic behavioral patterns of individuals particularly related to eating and physical activity. Using a multi-arm bandit approach (focusing on the suggestions that currently seem most promising), the system generated a list of concrete and actionable suggestions a person could engage in related to eating more healthfully or being more physically active. For example, the system would provide suggested routes that were slightly longer to enable incorporating physical activity into daily routines. The field is increasingly moving toward these more personalized behavior change interventions that are built using sensors either directly from the phone or augmented with other digital technologies. These examples are nice illustrations of a mash-up of data available from the phones themselves and augmented feedback systems.

Natural Environments and Complex Systems: A third research area is the study of natural environments and complex systems. On the environmental side, these include explorations into understanding how humans interact with their urban or work environments or the activities of animals within naturalistic environments. In terms of systems, a variety of industries are increasingly embedding sensors into their systems and processes to improve their management.

On the environmental side, much of these data are incorporated into the Global Information System (GIS). GIS includes an impressive array of layers of data that are geotagged such as names and types of businesses in specific locations, population density, average temperature and other weather data, and so on. These data are incorporated into GIS systems both via professionals and via crowd-sourcing strategies, such as pictures and stories of locations via micro-blogging options (Gaoncar et al., 2008) or walking, running, and cycling paths through a city (e.g., data from the

RunKeeper smartphone app). There are also numerous examples of this type of technology used for civic engagement, such as the study of how healthy different university campus areas appear to be in general (Griswold, 2004) or related to pollution levels (Estrin, 2014). Further, the Audubon society is a classic example of a crowd-sourced data stream for understanding bird migrations over time.

Beyond these scientist tasks, an increasingly wide range of industries are utilizing embedded sensors within context to better manage their systems. From supply chain management, to transportation (e.g., Hull et al., 2006), to aerospace and aviation, and even agriculture and environmental disaster management, embedded sensors are increasingly enabling a far richer data-driven understanding of the contexts in which we spend our time, with many of these data available via GIS or related API options. This is important for mobile user research as it establishes not only an interesting area of inquiry (i.e., understanding the movement of humans in place) but also provides added layers of information that can be used to better contextualize when, where, and often even with whom mobile apps are being used and not used. These data are particularly valuable when complemented with logging data, discussed below.

2.3   FACTORS TO TAKE INTO ACCOUNT WHEN SELECTING DATA SOURCES

In general, selection of sensors is dictated by the research questions of a particular study or project. That said, with any mobile user research question, there are often tradeoffs between accuracy and precision, usability, ethical practice, and generalizability of findings. We discuss each of these in turn.

Accuracy and Precision: Whenever a concept is inferred from a sensor, there exists the need to examine how accurately and precisely it measures the concept. *Accuracy* (also called validity) involves the degree to which an assessment is actually measuring what it is intended to measure. *Precision* (also called reliability) involves the degree to which the assessments can be replicated both over time within a single individual and across individuals. For some concepts, a "gold-standard" measurement strategy is available as a comparator for judging precision and accuracy against. This is often the case within highly regulated industries such as the automotive industry (e.g., there are exact standards for defining concepts such as speed or amount of gas consumed per distance traveled) or the healthcare industry (e.g., glucose estimates assessed via finger pricks). When such gold-standard tools exist and can fit into your mobile user research project (e.g., not excessively burdensome for participants), it is often wise to utilize the most cost-effective option available for sensing and inferring your desired variable.

Unfortunately, however, for many sensing strategies, no standards exist, or if they do, they may undermine the possibilities of the new devices. For example, while there are gold-standard metrics for assessing blood pressure using automated cuffs, an emerging area of research is the continuous assessment of blood pressure within context that does not require the use of a cuff. This is such a new strategy that there is no clear "gold standard" yet available for use as a comparator for continuous blood pressure assessment. This is because new information (i.e., continuous blood pressure) is available in the new devices that was not previously assessed with the old gold standard estimates. That said, clinical guidelines and other pragmatic decisions (e.g., "normal" blood pressure) ranges were defined using these older technologies and thus, it is unclear if "normal," for example, along with other key metrics and standards for interpreting these data will be the same when assessed using a continuous blood pressure monitor. At this moment in mobile user research, the distinctions on accuracy and precision, particularly on what is gained and lost when moving to a mobile user context compared to more "clinic" or "lab" based settings, is important to remain mindful of.

As a second example, there is a great deal of work currently taking place that's focused on attempting to assess stress nearly continuously and in context. Even within highly controlled lab environments, the concept of "stress" has been a difficult one to measure passively with a high degree of accuracy and precision. The classic, lab-based approaches for inferring stress often involve using a variety of sensors, particularly galvanic skin response, electrocardiograms for inferring heart rate and heart rate variability, and also respiration rates. While some mobile user tools infer stress through these sensors (though often just subsets of these) a wide range of other strategies for inferring "stress" have been explored including perceived "stress" based on fluctuations in a person's voice recorded via audio files (e.g., StressSense (Lu et al, 2012)) or the use of fewer sensors such as heart rate via LED diode sensors and other streams (Ertin et al., 2011). Often, these strategies are validated against self-report measures of stress, such as rating scales measured via the phone, thus establishing a relatively noisy standard as a starting metric. This results in several different "sensors" all labeled as measuring stress (e.g., Galvanic skin response, heart rate, audio files) but, often their level of precision, particularly the measurement of the concepts across individuals, and accuracy in terms of measuring a "gold standard" of stress are suspect. Further, it is essential to remain mindful that many of these technologies are currently being designed and deployed in relatively controlled environments (e.g., indoor quiet rooms). It is quite likely that many of the inference strategies will thus not work in a wide range of areas and conditions, a point that is particularly important for a mobile user researcher. Further, many of these emerging technologies, such as social sensing, utilize machine-learning techniques and are trained using human coders of the data to define "ground truth." Based on this, many inferences made about targets like movement or focus are only as good as the quality of humans' abilities to code these data from audio and/or visual data (Mast et al. 2015), and also need to take into account the frequent errors that may be produced by machine-learned classifiers.

These issues of accuracy and precision do not only exist with "sensors" but also within digital traces. For example, Tausczik and Pennebaker developed a wide range of different "dictionaries" for counting parts of written text, such as counting the types of words (e.g., verbs, nouns), grammatical issues with sentences (e.g., phrases, sentence structures), and the use of various "functional" words, such as pronouns within the system called, LIWC. The LIWC dictionaries have been shown to correlate with a wide range of psychological concepts such as personality, social hierarchy, and mood of the writer (Tausczik and Pennebaker, 2010). While these inferences can be valuable, it is important to remember that these patterns within free text are often, at best, simply correlations of other psychological patterns and thus may not be fully "measuring" the concept but, instead, merely providing an indication toward one personality type (e.g., neurotic, extroverted) or not.

Taken as a whole, the accuracy and precision of any inferred concept is essential to be mindful of when selecting a measurement strategy. In the "practical suggestions" section of this chapter, we will delineate some strategies for thinking through these issues in more detail.

Usability: One of the distinct advantages of conducting mobile user research is the opportunity for studying phenomena in the wild. This is particularly true when it comes to the integration of smartphone, digital trace and contextually embedded sensors as they provide a truly rich understanding of user's behaviors. Each sensor type has its own unique usability concerns to be mindful of though.

Smartphones are a highly alluring device for mobile user research because of how pervasive, personal, and powerful they are, with nearly two-thirds of the U.S. population owning a smartphone and having it near them at many times of the day (Smith, 2015). While there are advantages to the use of only a smartphone as a tool for gathering the myriad types of data that it can glean, as discussed above, a major problem with mobile user research often comes from assumptions about how a user actually uses a device. For example, a common assumption made by mobile user researchers is that a phone will always be on the person's body. If you feel comfortable making this assumption, it opens interesting opportunities for tracking physical activity using only the phone (Hekler et al. 2015), understanding movement within context around a community (e.g., the MyBehavior or RunKeeper phone apps), and understanding behaviors within indoor locations via proximity sensors.

As numerous personal examples of this will attest though, along with common sense, there are large individual differences in terms of how individuals often carry their phones. For example, many people (women in particular) carry their phones in a bag. This means that while the phone may be nearby, it is not necessarily on them, which makes accurate location tracking indoors via proximity sensors or activity tracking very difficult. Dey et al. (2011) found that smartphones are only at arm's reach 50% of the time, while 90% of the time they are within the same room as the user. This is fundamentally a socio-technical issue. The phones have the technical capabilities to sense a wide range of phenomena with the built-in sensors, but just because the phone can feasibly sense it does not mean that the participants in a study will use the phone according to your assumptions. We have had plenty of examples of this within our own work.

For example, in one physical activity evaluation study (Hekler et al. 2015) participants had large variations on how they would carry their phone, even when they were prescribed highly restrictive behaviors of either carrying it on a belt clip or in a pocket (this was for individuals aged 45+, who, based on earlier interviews with them, claimed to be willing to wear a phone on the clip or carried in a pocket). The algorithms that we generated explicitly compromised on precision to increase usability by allowing individuals to wear the phone in one of these two locations. Even with that, there was a fair amount of missing data, which was largely due to individuals forgetting to (or not desiring to) always carry the phone in their pockets or on their belts.

This problem became particularly pronounced for some individuals the longer we asked them to take part in the study. In particular, during the MILES trial (which utilized this sensor tool), our initial request was for participants to use the phones for only an eight-week period and then they were invited to use the phone for up to a year afterward to study more naturalistic behavior. One of the primary reasons why individuals declined to continue to use the MILES intervention app was because of wearability concerns with always having the phone on their person. With that said, this was highly idiosyncratic. Some individuals loved that they didn't need to carry any other device such as a pedometer or wrist activity monitor as, for them, carrying the phone with them during the day was completely natural. Our three-month field study of UbiFit found similar issues, in that sometimes participants did not have a place to holster the hip-mounted sensor kit (e.g., when wearing a dress) or simply did not want to wear it because of the attention it might draw (e.g., when dressed for a night out or an important business meeting). Overall, these examples highlight the need to carefully understand your target participants' natural behavioral patterns to ensure that you either: a) provide some value back to them if you are asking them to change their normal routines; or b) be OK with missing data from some potentially key user groups who do not use the phone as you had intended it to be used.

Digital traces are often alluring because of their usability potential. Specifically, these data are gleaned from digital actions (e.g., interactions with email, social media, etc) that a person is already engaging in. For those individuals then, there is plausibly a rich and easily "usable" strategy for gleaning data from their activity patterns. With that said, the issues here, from a usability perspective, are more closely linked with the data stream being used. In particular, it is common that most feeds for digital traces, such as social media or message boards, are often utilized for specific purposes for individuals, such as when bored or when seeking help or advice. This means that the data from these sources could feasibly be biased toward gathering data about a person only when they are in restricted states, which can then compromise the data analyses and generalizability (as discussed below).

For example, we conducted a study examining how different utterances (as coded using the LIWC dictionaries) within an online weight loss forum are associated with fluctuations in self-reported weight, which was reported using an accompanying app (Hekler et al. 2014). While the research team found interesting patterns that appeared to be correlated with changes in weight (e.g., use of past tense verbs, which tends to be indicative of an individual engaging in processing of past actions, appeared to be indicative of weight gain), there was also a considerable amount of missing data within this. It was more of the norm than the exception that individuals would participate in spurts on the discussion boards and also self-tracking within the app. As such, while there might be interesting patterns that emerged, the usability of the system, particularly during potentially interesting times to collect data for the researcher but not interesting for the user (e.g., if feeling ashamed about weight gain), compromises the research endeavor and the generalizability of any findings. Interestingly, you could feasibly incentivize participants to use a digital tool even when they normally do not use it but then this incentivization could easily influence the sorts of information provided (e.g., some LIWC dictionaries become biased or predict psychological phenomena differently depending on the prompts provided for writing information; see, for example, Tausczik and Pennebaker, 2010). As such, there are important usability concerns here as well.

Finally, there are particular usability struggles within the realm of wearable sensors. As the earlier phone example suggests, what a person actually carries around all of the time is highly personal and often highly idiosyncratic. A second key general trend though is that often some of the most precise and accurate devices are often some of the least usable or least attractive ones for individuals to wear, particularly over extended periods. Take, for example, the Zephyr Bioharness from Medtronic. It is a very well-validated chest-worn device that can be used to infer a wide range of information related to cardiovascular health, fitness, physical activity, and sleep, including assessment of issues such as sleep apnea, with numerous studies supporting its utility in both healthy and clinical populations. With that said, it requires relatively frequent charging (often at least once a day), it is often uncomfortable for individuals to wear over an extended period of time, is expensive to use, and also often is not particularly fashionable to wear, even with the band being hidden under a shirt. In contrast, something such as the Fitbit Charge HR, which is wrist-worn and can be accessorized with different bands and colors, is often more acceptable to wear by a wider range of individuals for longer durations but it lacks precision and accuracy in terms of tracking steps, physical activity, heart rate, and other factors that the Zephyr Bioharness does with precision and accuracy. This is a classic compromise between usability and accuracy/precision that often arises with wearable tech in particular.

When taken as a whole, issues of usability within mobile user research need to carefully take into consideration a wide range of idiosyncratic factors such as natural behavioral patterns prior to data collection, a person's sense of style and aesthetic, a person's varying perceptions of comfort for wearing devices, and the perceived motivations for a person to provide data. On the last point, for example, a patient diagnosed with heart failure is more likely to be motivated to wear a device with higher accuracy and precision than a person not diagnosed with heart failure. For them, the value of increased accuracy and precision could feasibly outweigh the added burdens. Any mobile user researcher looking to glean data from sensors needs to carefully map out and understand these behaviors of their target users prior to gathering data, otherwise, it is likely that the data will not be what was hoped for and could feasibly be meaningless. As an example of this failure, we conducted an exploratory study into the use of proximity sensors for providing just in time feedback for walking more within an office context. While we were aware of the socio-technical problem, we thought that we could circumvent this by making it a relative short study (four weeks) and incentivizing participation. Despite these efforts, our initial study failed to produce meaningful information about a person's movement within an office context (largely because so many individuals did not want to carry their phones with them) and, by extension, it provided limited information about just in time interventions. This example illustrates the potential usability challenges that must be constantly addressed within a mobile user research context.

Ethical Issues: There are serious privacy, security, and other ethical concerns that must be mitigated to ensure proper, ethical collection or use of any data. For example, one challenge from use of ecological microphone data is to be able to infer when a person is speaking while also ensuring the audio gathered cannot be recombined for inferring the specific content discussed between individuals (and thus breaching privacy) (Wyatt et al., 2011). Other ethical issues concern the need for robust informed consent to ensure an individual knows what they are actually signing up for and how their data will be used. The work of Sage Bionetworks and their consent process is a good example of a strategy that can be used to help ensure informed consent even in a digital context. The Sage strategy has individuals carefully walk through a series of questions and text statements to help them clearly understand what they are signing up for and to evaluate this understanding within the digital consent process. These ethical issues are important to take into account not only because it is the right thing to do but also because mobile user researchers are often interested in developing a more long-term relationship with their participants and target users, since the types of research questions explored with sensors are often particularly rich when there is a long temporal record of data. As such, careful examination of the ethical issues at the beginning can go a long way for supporting a more long-term sense of trust in the work and even open new avenues into better understanding phenomena.

For example, we are just starting to explore issues of researcher/participant relationships over time via one of our on-going studies focused on supporting physical activity behavior change. The focus of this line of inquiry is to develop highly personalized and precise behavior change interventions that are driven by individualized computational models that describe factors that influence each person's behavior (similar in spirit to Health Mashups) (Freigoun et al., 2017). For example, we've generated models that have shown that, for some individuals, when they self-report being more stressed they walk more, whereas for others, when they are more stressed, they walk less. And still for others, stress is not important but instead other factors such as if it is a weekday or weekend is more influential on their physical activity. Within our initial studies, we were very careful within our informed consent to ensure participants were aware of all of the highly sensitive data we wanted to collect about them such as location, calendar data, and interactions on social media, with each individual provided a way to opt out of any of these more sensitive data streams. Now that our individualized models are generated for each person, however, we are going back to our initial participants to inform them of what we learned. Our goal is to explore their reactions to the veracity of the models but also to invite them to become co-researchers with us in the process. Specifically, our next hope is that we can help them generate novel self-report questions that they will find interesting to track as potentially influential of their physical activity. Within this, the issues of doing ethical research have moved beyond just the mere idea of "doing no harm" to truly taking a more human-centered approach and actively engaging with our participants as partners in research.

Generalizability: Finally, a major issue with these new data streams are issues of access and thus generalizability. As discussed before, over two-thirds of the U.S. population currently have smartphones. This means that almost one-third of the U.S. population does not have smartphones, and many of those individuals may be among the more disadvantaged in our society. This same phenomena is true also for digital trace information and most definitely for contextually embedded sensors and wearable sensors, which tend to be expensive. What this means is that conducting this research could further perpetuate the study of a highly restricted range of human diversity, sometimes called the "WEIRD" types (i.e., western, educated, industrialized, rich, democratic). This issue, which has also been called the digital divide or information poverty, is essential to acknowledge and be mindful of within any mobile user research endeavor to ensure issues are not overly generalized. Your mobile user research questions might only be relevant for WEIRD individuals, but, it is important to remain mindful of how generalizable findings may be and to be clear about that as you report your findings. Before the smartphone received mass adoption, some researchers provided a new phone as part of the study incentive (Bentley and Metcalf, 2007; Ames and Naaman, 2007). This could be an appropriate strategy today for engaging a wider audience in field studies.

2.3.1   PRACTICAL SUGGESTIONS

With these four issues of accuracy/precision, usability, ethics, and generalizability delineated, we now turn to a few practical suggestions to be mindful of when using these data. Each is important to consider when planning your data collection mechanism and resulting study.

Understanding Your User Base

First, there is no compromise for robust ethnographic work for clearly understanding your target user base before, during, and after deployment of any systems to ensure you understand their natural behavioral patterns and routines and to also provide concrete grounding for carefully defining your research questions. These sensors are, in no way, a robust replacement for good ethnographic research methods and, indeed, often require even greater work to be done up front to ensure the data gathered is actually what the researcher intends it will be (as our proximity sensor study example illustrates).

Real Users Produce Real Surprises

Second, be prepared for idiosyncratic usage of your systems that you will likely not expect prior to deployment. We've discussed several examples of this already from different ways that people carry phones to different ways that individuals use online discussion boards. This can even influence how algorithms are developed and how generalizable and useful they are. For example, we conducted work on a variety of activity monitors and found that while these devices appeared appropriate for healthy adults, they were not necessarily as useful for older adults, particularly those with walking impairments, such as those who use canes (Floegel et al., 2016).

This problem can become worse with contextually embedded sensors and tools. For example, our work with BLE beacons has resulted in a wide range of responses such as, for some individuals, not knowing what the beacons look like and simply throwing them away, to others not recognizing the importance of them staying in a location and thus moving them to a location that they find more "aesthetically pleasing," thus compromising data collection. We have experienced these and many other issues that we did not anticipate when conducting mobile user research. Based on this, it is important that you are well-prepared to handle as many unexpected uses of your sensing deployments as possible. This often means that you end up being your own IT department for problem-solving issues that first arise in the field. To try and prepare for this, we strongly urge the use of pilot deployments with only a small group of target participants in a relatively controlled setting that would allow you to observe how your sensors and technology might actually be used in a real-world context. This can provide you with a wealth of information on what the data streams you collect actually mean, not just what you think they mean. As an example of this, we once conducted a study for understanding conference participant interactions using BLE beacons. We set up the space to ensure full coverage of the conference space, including outdoor tables where lunch was supposed to be held. Unfortunately, it rained that day and so the staff took all of the tables in, removed several, and then haphazardly moved all of our BLE beacons. This was only a small pilot, but it quickly taught us the importance of flexible planning (we had to reprogram and relabel our locations so that they were relevant and not overlapping) to ensure the signal we gathered was actually what we sought.

Ensure Data Collection is Usable

Third, assuming you have a good sense of your target group and a well-articulated research question, it is important to be mindful of the balance that often exists between accuracy/precision and usability. For example, let's assume that you are interested in using wearable data for measuring physical activity, particularly steps per day. If your research is focused more on the study of naturalistic behavior, you would likely want to use a wearable sensor/algorithm that has been shown useful for reliably inferring relatively accurate absolute levels of steps compared to a gold standard such as video-recorded step counts. For this type of question, it would become apparent that steps measured via a device worn on the waist are often more reliable at estimating steps across individuals and thus, would likely be more appropriate. That said, if the research question is more focused on 24-hour tracking of total activities, including sleep, waist-worn devices become problematic because they require a person to remember to transfer their device from their waist to their wrists at night, which often is forgotten (or to take the device off of the waistband of their pants, before they throw their pants in the hamper or washing machine). In this context, a balance between perhaps less reliable between-person estimates of absolute steps may be selected to increase the likelihood of gathering more accurate and useful data related to sleep. As a third plausible use case for step data, imagine your interest is more focused on fostering behavior change. In this scenario, you may find it acceptable to have less reliable estimates of steps between persons as long as you can have the data gathered for longer time periods, and thus you can feasibly measure and show changes in the outcome of steps. In this context, it might make more sense to use a wristworn activity monitor as they are often shown to be acceptable for measurement of steps over time within a person (i.e., any noise in absolute steps largely remains over time for an individual), thus making them acceptable for measuring changes over time.

Measure the Right Types of Data

Fourth, accuracy/precision must also be carefully thought through related to the research question to ensure what is "accurate" is actually appropriate for your research question. For example, the selection of the right "stress" measure from the ones discussed earlier is going to be greatly impacted by the research question. In situ estimates of stress via audio files might be particularly valuable for understanding stress within a social context but might be less useful for inferring stress during work activities, such as working on a computer when a person is not engaged in any major utterances and thus there is no clear audio-related signal of stress.

Consider Shared Devices

Fifth, related to usage data, while you might think that seeing all of a user's interactions with your application or system can tell you everything you need to know about how it is being used, it's important to consider limitations of this method. Phones and tablets are often shared with others (Matthews et al., 2016). It's not even unusual for people to loan their personal fitness devices to others (especially to children for brief periods of time who want to see how something works, e.g., to run around with mom's or dad's pedometer to watch the step count rise or to try to get the step count to rise by shaking the device in their hand). Usage logs from a single device might not directly map to usage for a single specific user.

In a recent study at Yahoo Labs (Holz, 2015), we were using an app logger to understand how participants were using their mobile devices while watching television. For one participant, we began to see what looked to us like odd usage partway through the study, with increased app use late at night and a different set of applications being used. In the final interview, we asked the participant about what we observed. The participant admitted to giving her phone to her mother for a week-long trip to China. These usage logs did not represent the activities of our participant, but rather those of her mother who was thousands of miles away at the time. We had to throw away this data.

Augment Logs with Qualitative Methods

Perhaps the biggest limitation of relying on log data is that it can only help reveal *what* people are doing, and *when* they are doing it. It does not give you insights into *why* they are doing it or if it's meeting their needs. For example, in Figure 2.1 above, you can see that features such as "logging pain" or "viewing graphs" were not frequently used. Combining the log data with open-ended questions in the final interview helped us understand how the pain feature was critical for those with chronic pain and simply was not needed by others. We also were able to understand how the graphs were not needed given the other data that was presented to participants. We discovered these important nuances through the interviews, such as more technically trained participants wanting to see the data in a graph before they believed a statement that we made about their well-being in a natural language sentence such as "You are happier on sunny days." Combining the logs with interview and voicemail data gave us a much more full understanding about what was occurring.

2.4   CONCLUSION

Mobile sensing and the collection of usage data provide great opportunities for creating novel applications, systems, and studies that go beyond data collection that was possible with previous devices. From on-device sensing such as accelerometers and GPS to wearable devices and environmental or other Internet of Things-based sensors, there is a vast set of data that can be collected.

However, collecting this data ethically, in a privacy- and security-observant manner, and accurately inferring higher-level behavior from the data remain topics with many open questions. While this chapter has shown some of the opportunities for using this rich data, it has also highlighted some of the issues to address when designing your system and data collection procedures. These topics will be important as you move to later chapters in this book. You now have a toolkit of data that can be collected and analyzed, and the later chapters will explore how to use this data to answer a wide variety of research questions through both controlled and in-the-wild experiments.

CHAPTER 3

Observations in the Field and in the Lab

3.1 INTRODUCTION

In order to create a system that fits into everyday life, it is extremely important to learn from a wide range of people at all stages of research and development. You and your research or design team come to the table with certain perspectives, and these perspectives are often not those of a wider audience. "You are not the user" is a common mantra of HCI research. It generally holds true when particularly tech-savvy professionals are creating products for use in a wider user base that does not have the same background or intimate knowledge of the technology being developed. Even when you consider yourself to be a member of the audience that you are designing for, you'll often be surprised by the range of behavior that you can observe from a sample of others in your target audience.

Before even settling on an idea or initial design, *exploratory* (also called *generative* or *formative*) field studies can help you understand current attitudes, practices, and needs in a given domain. Such an understanding can lead to countless ideas for the development of new systems or features to better meet people's needs.

As you begin to develop a solution, iterative rounds of evaluative research in the field can help you understand how potential users will understand and engage with your solution. These studies can involve "Wizard of Oz" experiments where researchers fake certain parts of a system that are not yet built (discussed further below). Through these studies, you can begin to learn how people would engage with your system if it was fully built, and you can make major strategic changes without having invested large amounts of time on system development.

Finally, before shipping a product or running a long-term field study of a new research system, it is important to have external participants engage with the system to discover usability issues and to refine the design to be understandable by a wide range of people. Many parts of this step can be conducted in a lab, and we will describe some best practices for this type of research.

When system creators do not follow these processes, unintended consequences can occur (and, unfortunately even when they do follow these processes, unintended consequences occur, but the risk is reduced). For example, *Formspring* was a social network heavily used by teens that ended up attracting a large amount of bullying and hateful content (Watkins, 2013). The team focused on following industry trends over designing features that their users actually needed, and they hung on to an anonymous model that continued to promote hateful speech. This ultimately led to the closure of the service. By repeatedly testing concepts with users in real-world environments, discovering "unintended" uses, and rapidly redesigning to meet user needs, products can anticipate negative uses and correct their direction before unexpected uses end up spiraling out of control, potentially in a more public context.

Historically, not exploring use in real-world situations has led to medical errors and even deaths. The Therac-25 system (Leveson and Turner, 1993) was a radiation dosing system that was prone to frequent malfunctions. Operators learned to ignore errors—many of which were unimportant—and when combined with software bugs, several deaths, and radiation overdoses resulted. As the system developers were not aware of the real-world usage and on-the-job training that new operators received (where operators ignored errors or used the console in ways that it was not designed), they had not anticipated the conditions that led to the tragic overdoses, and it was difficult to understand how the system was failing in practice without having this real-world observation of use.

Overall, it is important to engage with participants who are representative of your target audience (and not part of your research team) iteratively throughout your project—use the cheapest reasonable approach to get the type of information or feedback that you need, as early and often as possible. You can learn from your target audience's current practices and from their feedback on early concepts so that you can quickly change course before spending large amounts of time designing and writing code for solutions that will not fit into people's lives or are difficult to understand and use. Once you develop your technology, study it in the field to make sure that it works as well as you hoped, and keep an open mind that's ready to make changes when you learn that something isn't working well (even if you spent a lot of time and effort developing it and you think it's amazing).

3.2   EXPLORATORY FIELD STUDIES

It's best to conduct exploratory (also called generative or formative) field studies early in a project, before the idea of what you want to make is even fleshed out (it's okay if you have some initial ideas—just don't be too wedded to them yet). Since mobile technologies will often be used where people live, work, and/or spend their free time, it is important to visit these places and understand people's current attitudes, practices, and needs before embarking on a new project or finalizing a design. Doing so will ground your new idea in people's real-world behaviors and can help to ensure that you'll create something that people (and not just you or the research team) will want and be able to use.

When beginning to design, many questions often arise about the places where the technology will be used. Seeing and understanding a study participant's home, car, or workplace, and how they interact with others in those places, can often provide needed design inspiration to develop a new idea that truly fits into people's lives. Also, while evaluating new mobile concepts, it's often helpful to return to the field, visiting your study participants to see the types of places where they use your system. Seeing the environment will make interpreting other data that you might collect (e.g., through a *diary study*—see Chapter 4) easier as you'll be familiar with the places that participants mention and the tasks that they talked about when you visited them.

Field visits have a strong history in *Anthropology* and *Ethnography*, where scientists have been visiting homes, studying the artifacts of environments, and understanding how objects are used in daily life for decades. While anthropologists and ethnographers traditionally use deep ethnographic methods, involving months or even years of living around and interacting with their participants, faster methods have emerged. Beebe, in his *Rapid Assessment Method* (2001, 2005), explored ways to use semi-structured interviews as well as semi-structured site visits to quickly understand practices of interest. These faster methods sought to triangulate a variety of methods to arrive at an understanding of a given practice. Visits to homes and other sites of interest became a key component of this method. *Triangulation* uses multiple methods (e.g., interviews, diaries, observations, quantitative logs, etc.) to gain different perspectives on a practice of interest. Each of these methods can provide data that is missing in other methods, providing a larger, more well-rounded picture of the practices of interest. This is ideally how site visits, such as *home tours*, can be used in field research (i.e., to provide background and contextual details that are not often possible to collect when only using interview-based or quantitative methods).

Home tours are also closely related to an HCI method known as *Contextual Inquiry* (mentioned in Chapter 1). Developed by Hugh Beyer and Karen Holtzblatt (1998), this method involves going to the places where people perform tasks and asking them to perform that task, in its natural context. This idea of traveling with users to contexts of typical interactions heavily influenced later HCI researchers who began to regularly incorporate home tours into their methods. This became especially important as computing moved away from fixed desktop computers to portable and mobile devices that can be used in a wider variety of contexts. Mobile user research, almost by definition, requires studying technology in everyday contexts of use (and not just in the lab).

3.2.1   FIELD STUDY TIPS

Planning a study that involves locations in the field is not difficult, but it does involve advanced planning. If you have a method including some sort of longitudinal data collection (such as the *diary studies* described in Chapter 4), field visits are best conducted at the start of a study, often as part of the *initial interview*, so that you can use what you learn from the visit to help interpret the data that you might gain through other study components, such as diaries or *usage logs* (discussed in Chapter 2). When planning your study, think about your research questions and the types of places that are most important to see. For example, if you are studying how people communicate, make sure to ask about recent places where participants have placed phone calls, video chats, or written letters. If you are interested in music use, ask to see all of the speakers in the home, or places where they might sit and listen to music through headphones. In this case, you might also want to see their car, where they exercise, or other venues where they like to listen to music. Make a list of these places and the questions you'd like to explore in each location. Often, the best questions simply consist of asking the participant to tell you the story of the last time they performed the task that you are interested in while you're together with them in the place they performed that task. Be sure to ask follow-up questions to get the relevant details of these occurrences. In addition, you can use methods from Contextual Inquiry to have participants demonstrate the interaction in the place where it occurred so that you can see how the environment may have affected the task.

Before you visit a participant's home or other environment in the field, there are a few practical details to consider. First, make sure to tell the participant that they should not clean or otherwise alter the space (e.g., their home) before you visit. They likely will do a little cleaning regardless. However, if you make it clear to them that if they clean up, it might make it harder to do your research, they might be more inclined to keep things close to normal.

A field visit usually begins with a short, *semi-structured interview* (briefly discussed in Chapter 1) with the participant to help the researchers understand a bit more about their lives and to build trust between the researchers and participant. This interview can focus on topics of interest, for example, asking participants to discuss recent times when they performed tasks of interest to the research questions at hand. Following this interview is an opportune time for a home tour as the participants will feel more comfortable talking with the research team and sharing more details about their lives.

We generally video record our home tours—with the participants' permission—so that we can later recall any visual details about the environment that are interesting as additional data becomes available. One researcher generally records while the other leads the interview. Participants are asked to take researchers to the specific places where recent interactions have occurred and to discuss the context around those actions. As discussed in more detail below, sometimes participants can be asked to re-enact particular tasks or to show the researchers how they would perform a similar task in that location. This helps the researchers to more fully understand how the context of that particular place plays into the participant's actions. The research team can then ask follow-up questions to clarify the details of any particular action, or ask about other objects in the environment (e.g., asking about recent interactions with an otherwise unmentioned stack of CDs by the stereo in a study about music).

It is important to note that places like the home may be shared environments. While you are there, other occupants might want to join in the interview or appear in your video recording. Research teams should have a plan for these situations in advance and have approved them with their Institutional Review Board (or related reviewers). For example, it might be necessary to bring extra consent forms for others to sign. Or it might not be possible to involve minors and they might have to be asked to leave the area during filming. It is also important to ask participants in advance if there is anything you should not film (e.g., a piece of mail that contains personally identifiable information), and to respect their wishes in this regard. You should be ready to edit the recording to delete something that was captured but shouldn't have been (even if you only found out that it shouldn't have been recorded after the fact).

After the tour ends and you are back in the lab, several types of analyses are possible. The audio from the interviews and home tours can be transcribed and analyzed using *open coding* (Burnard, 1991) or a *grounded theory-based affinity* (Birks and Mills, 2011).[11] In a more visual type of analysis, videos can be watched, and key incidents can be noted. Maps of the home can also be drawn to help in interpreting future data to be collected via voicemails or automated logs if these methods are to be used in parallel. Data collected from generative field research can be used in many ways, and we often find ourselves returning to videotapes or notes from a field visit throughout the analysis process. We recommend some of our go-to resources for how to analyze qualitative data in Chapter 1.

Variations

Many variations to the field study method exist. The first two that we discuss—the self-guided video tour and live video interview—are useful when it is not possible for the researchers to visit the field location in person, for example, if you have recruited participants from across a country or from other parts of the world. The second two—shadowing and contextual inquiry—are ways to augment a field visit to collect even more detail.

When it's not possible for researchers to physically visit participants in the field, you can ask participants to provide a *self-guided video tour*. In these instances, they can record a video of their environments, highlighting places and topics that you are interested in learning about (you may need to provide them with recording equipment). This can be helpful if you want to see different parts of their lives. For example, they could capture a video or series of videos while shopping or going for a run to show you more details of these environments. Or if you're studying travelers, they could capture videos of their hotel rooms or other places where it would be uncomfortable or inappropriate for the researchers to visit. This technique could be used in place of or to augment the type of field visit we discussed above.

Another option is a *live video interview*, which can be conducted with video conferencing equipment (e.g., Skype or Google Hangouts should be easy for your participants to access). With

---

[11] We used Lag Sequential Analysis to analyse video taken of biologists working in their laboratory (Consolvo, et al., 2002).

this method, you can guide a participant around their space, asking for more details about particular objects or areas, similar to what you could do in an in-person interview. Unfortunately, much of the detail is often lost with live video, and you also lose the chance to look around the environment yourself to find other potential objects of interest that the participant might not explicitly show you. However, this is almost always better than nothing in learning a bit about a remote participant's context. It can also be used if participants are from different corners of the world where travel might be prohibitively expensive. You might also want to use both of these methods; for example, after using one of these methods, you could follow-up with the other to fill in missing details from the first method.

Going to the other extreme, there are times when you might want to follow someone into a variety of environments throughout the day, otherwise known as *shadowing*. Metcalf and Harboe (2006) were interested in exploring how people managed boundaries between work and home. To uncover this, they shadowed diverse participants throughout the day to see how home-focused tasks were handled at work and work-focused tasks crept into home life. While a diary (see Chapter 4) might have captured some of these events, actually seeing people juggle these diverse aspects of their lives brought increased detail and understanding to the data as well as built a deeper level of rapport with the participants.

Field visits can be combined with other methods such as Contextual Inquiry (Beyer and Holtzblatt, 1998) to see how tasks are performed in different rooms of the home. For example, if you are studying music selection and use, a participant might be asked to play music as if they were having friends over while they show you the living room, or to play music in their office that they would listen to while working. Watching participants perform tasks where they would normally perform those tasks can provide deep insights for new technology concepts based on places where participants get stuck or need to improvise using today's tools.

Finally, field visits might be the start to a larger study that involves participant data collection (such as a diary study) or the use of a system in the wild (also known as *evaluative field studies*, which are discussed later in this chapter). In these instances, the field visit can give researchers additional context that can be very helpful for interpreting data collected through other methods as the study continues. It is also a chance to build rapport with participants, which often leads to richer data reporting in other methods used in the study.

Limitations

While field visits can provide a great amount of additional data and richness about a person's environment, there are limitations to this method. First, you are only getting to see the environment at a snapshot in time. Places, such as homes, are dynamic spaces that change throughout the day. The kitchen table can look very different at dinnertime compared to when kids have it covered with homework or when it's being used to make cookies. Likewise, the family room can look very different in the middle of the day compared to during a Super Bowl party. While seeing the environment and asking people to remember tasks that they performed in different rooms can get you some interesting data, it's not unusual for participants to forget to mention (or choose not to mention) certain uses of the room. To counteract this somewhat, researchers can take care to ask about other uses of rooms, and if appropriate, ask participants to reconfigure a space as they might for certain activities. However, they can't tell you what they don't know about. If the room is used in other ways by other household members, they may not even know to tell you about that.

No matter how much you ask participants not to clean, they likely will to some degree. Thus, you will be seeing a vision of their space that's closer to what they would like to project to guests, and not necessarily the one that they live in from day to day. Piles of bills might be tidied up and put away, stray clothes may be in the hamper and not draped on the sofa or in a pile on the floor, and children's toys may be neatly put away in the toy box. If you've built enough rapport with your participants and have a feeling that items of interest to your research are missing, you can often ask them explicitly about these topics and what they might have done to prepare.

Finally, you are relying on the participant's recall of the actions that they performed in particular places. Recall often misses, or misremembers, key details of interactions. Combining a field visit with methods such as diary studies (see Chapter 4) can help you get better insight into what participants *actually* do in particular spaces, instead of their often-idealized memories of events.

3.2.2   EXAMPLES OF GENERATIVE FIELD STUDIES

In this section, we share two specific examples of how we used field studies as part of larger generative research programs to help us understand how one's environment affected their communication (e.g., the *Elder Communication Study*) or their use of music in the home (e.g., the *Music Context Study*).

Example: Elder Communications Study

At Motorola Labs, we were interested in developing a new communications application to bring together family across generations and distance. To begin this work, we conducted a study of current communications practices between young adults (aged 18–34) from the greater Chicago area and their parents or grandparents who had retired to Florida (Bentley and Basapur, 2012). The study used a variety of methods to understand current practices. We conducted an initial interview in participants' homes to discuss recent episodes of communication. During this interview, we asked participants to give us a tour of their homes to see places where they communicated, the devices they use, and any artifacts they kept around the home to remember their family members who lived at a distance. Then, for the next three weeks, we asked them to keep a voicemail diary of any communication that they had with the family member of interest. Finally, we conducted another interview in their home where we asked follow-up questions from the voicemail diary. Together, these methods gave us a deep picture of their communication routine.

The home tour was conducted during the initial home visit, just after the initial interview (which was usually conducted at their kitchen table). This allowed us to build some rapport with the participants before asking to see more private areas of their home. The visit was always conducted with two researchers, and we tried to have one male and one female researcher at each visit. We began the home tour in the same room as the interview, asking participants if they remembered the last time that they had communicated in that space. After discussing recent communications, we asked about artifacts of remembrance that were placed in that room (e.g., photos, cards, letters, etc.). We asked how long it had been there and why they chose that particular artifact and location. We then moved around the house, asking the same questions in each room. Often, interviews took us to backyards or cars as those areas were also frequent places where participants communicated with others.

> **Tip:** Don't limit yourself to the physical structure of the home if the participant is comfortable sharing more. But, of course, do respect their boundaries and don't pressure them into showing you something they're not comfortable showing you or don't have permission to show you, e.g., the room of a sibling.

As we received data during the next three weeks via voicemail diaries, we could more easily interpret the context of the participant. They knew that we had seen their home, so they would tell us things like "*I called my mom tonight from the chair out on the patio*" and we'd know what they meant and could picture them sitting there on the phone. As we had video recorded the tours, we returned to see particular places in the video during our analysis and transcribed the events discussed during the tours for inclusion in our larger qualitative data analysis sessions from data obtained throughout the study.

In the end, *place* played a large role in the solution that we developed, *StoryPlace.me* (Bentley and Basapur, 2012) (see Figure 3.1). However, we moved beyond the home to create a system that allowed people to tell stories about particular places in the world that were important to them and share them with their family members. This was directly inspired by data obtained during the study.

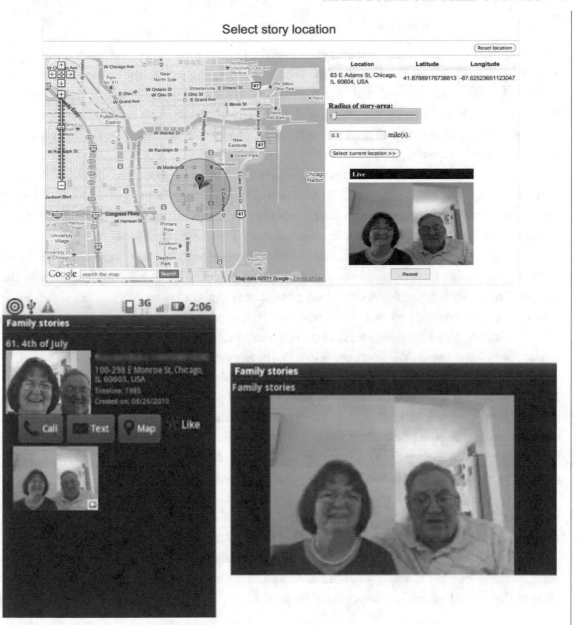

Figure 3.1: The Serendipitous Family Stories system, which was the initial prototype that led to the StoryPlace.me system. The system we created was a direct result of our elder communications study. The system allowed parents and grandparents to leave stories for their children and grandchildren in different locations in the city. We used findings about the importance of place and of family history in maintaining family bonds over a distance in the design process. From Bentley et al. (2011).

Example: Music Context Study

In 2004, the world of music was changing rapidly. Digital music and downloads were beginning to replace CDs and fixed stereos. The iPod was promising your entire music library in your pocket. At Motorola, we predicted that the music player would soon be part of the cell phone. But instead of just adopting the album and artist lists of the iPod, we wanted to use existing music practices to inspire us to create something totally new for the digital world that addressed people's desired practices.

To investigate how music fit into people's lives, we visited the homes of diverse participants for a home tour that was combined with a Contextual Inquiry study (Bentley et al., 2006). We were particularly interested in why specific music was kept in certain rooms of the home (generally in physical CD form) and how participants selected music for specific occasions.

As in the *Elder Communications Study* discussed above, we began our home tour with an interview, though this time we asked about general music practices. We asked participants about the last few times they listened to music and the context around those events. We also asked for details about how they acquired new music and how they learned about new bands.

Following this interview, we asked participants to take us to the "main" place in their home where they played music. We asked them to show us any music that was in that location as well as typical scenarios for playing music in that location. At this point, we changed from a traditional home tour toward more of a contextual inquiry. We asked participants to act out one of the music selection-and-playback tasks that they had discussed during the initial interview. Examples included putting on some music for studying, for inviting friends over for dinner, or selecting music for a road trip. While they did this, we asked them to "think aloud" (Nielsen, 1993; Nielsen, 2012) and tell us what was going through their minds as they completed the task.

By using this method, we learned about the importance of music that's "available" and in the same room as the participant. This made us think about new interfaces for having related music available at one click in a user interface and the "playtree" concept (see Figure 3.2) (Harboe et al., 2010). The playtree concept showed music that was related in some way to the currently playing song, offering multiple branching options at any point during music playback (with the default of playing the current album/playlist straight through).

Figure 3.2: The "Playtree" concept. Options for the next song are shown based on particular metadata aspects of the currently playing song (e.g., Genre, Tempo, Year Published, etc.). From Harboe et al. (2010).

Figure 3.3: The Music Dial as later implemented on a smartwatch. The user could turn the dial to select music published in other years, or swipe to the side to switch to selecting based on tempo, or recently/frequently played music, saving the need for scrolling through large lists on such a small screen. Courtesy of Motorola, Inc.

We also saw how participants often "satisficed"[12] to hone in on music they would enjoy listening to. Frequently, they tried to find music that was slower or faster than the currently playing song, or perhaps something newer or older. This led us to create the "Music Dial" concept (see Figure 3.3) (Bentley et al., 2013a) that allowed for music selection to take place simply by spinning a dial

---

[12]   "Satisficing" is a term coined by Herbert Simon that describes a decision-making strategy where people aim to satisfy the minimum requirements necessary (Simon, 1956).

tied to beats per minute (BPM) or year published. Similar knobs could be constructed that allowed participants to select music that they frequently play, or music that they have not heard in a long time. These ideas came directly from observations in the homes of participants and watching them interact with their own music collections in their natural contexts.

3.3   EVALUATIVE FIELD STUDIES

While exploratory or generative field studies can help you learn a lot about the context in which your technology will be used, it is often necessary to study *how* your system is actually used once it's out in the field. Evaluative field studies can show how often users engage with your system as well as the ways that it fits (or doesn't fit) into their routines and contexts of use. As you prepare to take a new system to market, or if you want to test how a new research concept can alter daily life, an evaluative field study is often the best route to take before a full public release.

Evaluative field studies often combine many of the methods that are covered elsewhere in this book including instrumentation/logging (Chapter 2), diaries (Chapter 4), or experience sampling (Chapter 4), as well as some methods that are mentioned in Chapter 1 (e.g., interviews and surveys), and comparative trial designs, if their is a clear focus on changing behavior (Chapter 5[13]). They generally occur over a minimum period of three or more weeks,[14] with studies focused on measuring behavior change often lasting several months or longer. Because of this, there are many practical considerations that must be addressed before starting an evaluative field study.

3.3.1   PREPARING FOR AN EVALUATIVE FIELD STUDY

Before you put your technology in the field—even for a study—there are many issues to consider in preparing your technology for use out in the world by people who are not members of the research team. Some of these are very practical and involve hardening your technology for real-world conditions, where hackers may try to break into your service. For example, you should ensure that any server-side components use a secure communication protocol such as HTTPS, *properly* SALT and hash passwords, and *properly* clean any inputs for possible injections. Think like a hacker trying to break into your service—where are the vulnerabilities? Come up with a plan for when and how you will delete the data you collect. Consult computer security and legal experts to make sure you're doing the right thing—this is your responsibility. Your study participants will often be trusting you

---

[13]   Note that different fields often require different types of evidence for making definitive claims. For example, if you are working on a health-related mobile interaction, the expectation is often still that you run a large-scale randomized controlled trial. Be mindful of this when doing any evaluative trial.

[14]   In our experiences, evaluative field studies that are shorter than 3 weeks are typically not long enough to uncover important problems people will experience with the technology, in part due to novelty effects. This can leave the researchers with an unrealistically optimistic view of the technology and miss the opportunity to learn about important problems that need to be addressed before a wider or longer release.

with very sensitive data, possibly including their location, health data, personal communications, and even data from or about others depending on what you are evaluating, and you should do everything you can to keep this data safe and secure from others.

If your study involves having the participant wear, carry, or otherwise use some type of hardware (commercial or custom), there are other important considerations. Safety, comfort, and aesthetics matter. Your hardware also shouldn't interfere with their existing technologies. How well will your hardware work with their other technology and integrate with their environment? Will it work well with their style? Will it cause damage to their clothing or draw unwanted attention from others? Will it annoy others? Will it survive a fall or getting knocked around in a bag? What happens if they lose it? What if it falls in the toilet or ends up going through a cleaning cycle in the washing machine?[15] The longer you ask participants to wear, carry, or use something, the more important these considerations are.

In addition to the above, it is extremely important to test your system and your study protocol in daily use with a pilot group before giving it to unknown participants. We often try to get colleagues (especially those who aren't on the research team) or their family members or close friends to participate in pilot studies. Even these "friendly" pilot participants are bound to uncover issues you hadn't considered; fixing the problems uncovered during piloting before your "official" study will help you get much more out of the "official" study with unknown participants (why waste their time or yours uncovering known/easily addressable problems?). Make sure to actively test your technology in places without network coverage, enable generous retries to send data to your server, and securely cache as much as possible so that your system continues to work even if the network is unreliable. If your system involves any type of activity or place detection, make sure to rigorously test with a variety of different people, places, and transportation types before deploying to a broader audience. We have found that pilot testing should take place over weekdays *and* weekends to ensure that you have considered a broad range of situations, and for a minimum of five days. We've even piloted prototype systems for many months to work out the kinks before conducting any "official" studies (e.g., we used this approach during the development of our UbiFit system (Consolvo et al., 2014)).

In our experience, it's not unusual to take at least a month to get from our first working prototype to a reasonably hardened prototype that's ready for a field test, where we are confident that the system will work well enough in the daily lives of our participants to lead to a successful study. If your system is particularly complicated, requires custom hardware, or novel inference, it could take much longer. Taking less time than this to test and prepare may very well lead to unexpected issues with your participants and perhaps needing to throw away large amounts of data (or even all of it) when things do not work. Thinking about various networking scenarios and data formats

---

[15] In multiple studies, we have had participants report that a research prototype or a piece of commercially-available hardware that we loaned them ended up in the toilet or washing machine.

(e.g., if you're building a system that takes text input, supporting extended character sets for emojis not only in your app, but also in the server software and database as well) can help you move a long way toward a solid system. Most grad students or programmers early in their careers will generally need to budget several weeks or more of *additional* effort in getting an application to a solid base that can be used for weeks at a time by participants who are unfamiliar with the technology.

When starting the study, make sure that your participants know that they are trying out a very early version of a new concept and that there might be issues with the software or hardware. Let them know that any problems they encounter will not be their fault, but are likely issues in the very early technology. Provide a phone number or email address for always-available technical support during the study and be ready to help participants with issues that might arise. Using methods mentioned in other chapters of this book, for example voicemail diaries (Chapter 4), experience sampling (we offer an example of this later in Chapter 4), or instrumentation/app logging (Chapter 2), can help you to ensure that your system is working as designed throughout the study. If it has been two or three days since you were expecting to hear or see something from a participant, particularly if their usage logs suggest non-use, check in with them to make sure that everything is working.

Example: Serendipitous Family Stories

When we designed the Serendipitous Family Stories application at Motorola Labs (Bentley et al., 2011), we wanted to discover how people would use it in their daily lives. The system was created to promote communication by sharing and discovering family stories at places in the real world. It allowed people to capture videos and save them to a point on a map. They could then share these videos with friends or family who would uncover them as they walked near the location of a story. The recipient's phone would vibrate if a story was available nearby, and with a single click, they could watch the video with the sender talking about the importance of the place where they were currently standing in their family history.

To evaluate this concept, we ran a one-month field study with 20 participants. We recruited young adults in the Chicago area and their parents or grandparents who had grown up in Chicago but had now retired to South Florida. We met with the older generation in their homes where we provided a webcam and helped them to record stories about their lives growing up in Chicago to be shared with their children or grandchildren.

Then, for the next month, the Chicago participants used the system on their smartphones, receiving vibration-based notifications any time they came near a story. We asked participants to call a voicemail system in the evenings on days when they uncovered a new story to tell us about their experiences. The app also sent usage logs to our server each time a screen was viewed or a video was played. We could also track if participants called or text messaged the sender immediately after watching the video.

At the end of the study, we held semi-structured interviews with both the younger and older generations to understand their use of the system, ask follow up questions from the voicemail diaries, and ask about general communications patterns between the generations.

The qualitative data from interviews and voicemail diaries combined with the quantitative data from the app usage logs gave us a bigger picture about how this application was used and how it fit into the lives of our participants. We saw weekend days where the app was actively used to go "hunting for stories" in the city. We saw weekday events where stories were stumbled upon as participants went about their daily lives. Most importantly, we heard stories about how the system was helping to bridge the gap between generations. One participant told us about learning that her grandmother was not "just a sales clerk" at a department store, but was actually an executive with the corporate headquarters. This led to a multi-hour phone call about the role of women in large corporations, just as the younger participant was starting out on her own career. We also heard details about children discovering things that they never knew about their parents, like a love of architecture or stories about romantic dates that they used to go on. And we saw how the stories helped participants understand what the city was like decades ago, where an amusement park stood on the banks of the river and people took electric streetcars everywhere they wanted to go.

The rich combination of methods in this study helped us to get a broader picture of use out in the world. As we couldn't follow someone around at all times, usage logs combined with a voicemail diary got us as close as we could get to the action of uncovering a new story out in the world. As a result of the study, we realized the importance of having additional content in the system from others, so that there was something to consume when there was nothing nearby from a family member. We ended up partnering with a variety of PBS stations and tourism boards to provide public content in many cities throughout the country that users could follow when they first downloaded the app, before any content was explicitly shared with them. We also learned a lot about the experience of being notified about a story. The system initially had two different notifications—one when you were "close" to a story (within a half mile) and then another when you were actually at the location where you could unlock the video. Participants often came close to a story while they were commuting on the train, and would be bothered everyday with the notification with no way to get to the destination as the train was moving past it. In the next iteration, we enabled users to mute notifications for stories as a way to avoid this twice-a-day notification.

3.3.2   WIZARD OF OZ METHODS

Sometimes, you might have a research question that you would like to answer that would involve building a complex—or expensive—system to properly answer. At times, the technology to do so might not be currently available but you would like to see if a certain experience can be enjoyable or useful for potential users. A method known as *Wizard of Oz* (Wilson and Rosenberg, 1988;

Landauer, 1987) can enable researchers to fake various aspects of a system while providing the participant with an experience that makes it feel like the system is working.

Wizard of Oz is often used when researchers are unsure if it is worthwhile to build a system. They can test a participant's reaction to a "working" system by working "behind the curtain" to perform tasks that the technology would normally perform if the system were fully implemented. Before accurate speech recognition was available, this method was commonly used to gauge participants' reactions to the types of errors that systems would likely make; to do so, humans would rapidly type in a separate room from the one the participant was in, and the text would "magically" appear in front of the participant who was speaking (e.g., Gould et al., 1983). The method was also heavily used in computer vision research to perform tasks that were currently difficult for software to perform, for example Darrell et al.'s (2002) work on creating a *Look to Talk* system where a participant looked at a conversational agent to issue commands.

When using the Wizard of Oz method in the field, it is usually a good idea to keep your "Wizard" away from the participants, so that they do not see that part of the task is not being accomplished by the system. This might mean that you need to implement waiting dialogs or other graphical elements to display to the participant that plausibly explain the delay that might be imposed by inserting a human into the processing loop.

There are a wide variety of studies in the HCI literature that have used field studies to evaluate prototypes of varying fidelity. Next, we'll highlight two that used Wizard of Oz in the field, as part of larger research projects.

Example: Photo-Based Search

Tom Yeh and colleagues from MIT's Artificial Intelligence Lab created a Wizard of Oz system for visitors to the MIT campus (Yeh and Darrell, 2007). The ultimate goal of the project was to create a computer vision system that could identify landmarks around campus. Visitors could take photos on their mobile phone to learn more about these objects. However, many questions appeared that could guide the computer vision work. What types of photos would people take? Which features of buildings would be included? What did the long tail look like? And once the building/object was recognized, what type of information did visitors want?

The researchers built a Wizard of Oz system where the photos that participants took were sent to a researcher sitting at the ready at a computer. This researcher had preset responses ready to send for the most popular landmarks on campus. As soon as the researcher received a picture, they selected the appropriate landmark and replied with the relevant information. While this was slower than a real system would execute, researchers could then interview the participants about the usefulness of the data that was returned as well as collect a nice corpus of labeled images from which to train their computer vision system for the ultimate fully functional application.

Example: The CareNet Display

In the early 2000's at Intel Labs Seattle, we were working on developing technologies to help provide elders with the care they needed to remain at home. Prior work—Mynatt et al.'s *Digital Family Portrait* (Mynatt et al., 2001)—introduced the idea of using a digital picture frame augmented with information about the elder's day to offer peace of mind to *distant* family members who were concerned with the elder's well-being. The Digital Family Portrait was designed to be kept in the homes of the distant family members and updated with information about the elder that was detected by sensors in the elder's home. In our work, we were focused on supporting the *local* family and friends who provided the elder with care. The Digital Family Portrait inspired us to develop an interactive digital picture frame—the *CareNet Display*—that augmented a photo of an elder with information about their daily life that would help their family and friends provide their care (see Figure 3.4) (Consolvo et al., 2004a). Similar to the Digital Family Portrait, we intended that the information about the elder's daily life that was shown in the CareNet Display would largely be provided by sensors in the elder's home.

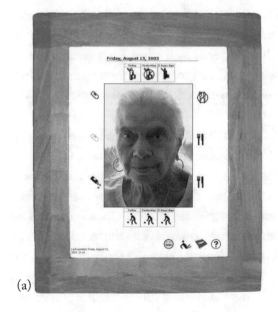
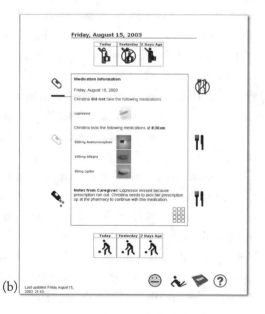

Figure 3.4: The CareNet Display prototype. (a) From the CareNet Display's main screen, users could get an overall picture of the elder's day from a glance, or they could interact with the touch display for details. (b) The "morning medication" detail screen. From Consolvo and Towle (2005).

In addition to being inspired by prior work, the CareNet Display was informed by two earlier studies that we conducted: a set of semi-structured interviews with various stakeholders (including elders and those who provided their care) (Consolvo et al., 2004b) and a set of roundtable discussions with various stakeholders to explore their reactions to possible technologies to support the elder's nonprofessional care providers (Roessler et al., 2004). Although we had already conducted two sets of studies to inform this work, we were still relatively early in the development of the sensing and inference technology that would be needed to support it. We also thought that we had a good idea with the CareNet Display, but weren't sure how the target user population (including the elders) would react to this rather unfamiliar technology that was communicating potentially sensitive information (e.g., how would the elder feel about this information being provided on a picture frame in a family member's home? How would this technology affect communication between the various stakeholders? Had we picked the right information to show?[16]).

To see if we were on the right track, we evaluated the CareNet Display in an in-situ Wizard-of-Oz deployment of our prototype. We built three identical prototypes of the CareNet Display to deploy in the homes of participants who were representative of our target users to see what impact it would have on elders and the local family members who cared for them. The deployments ran for three weeks at a time and involved the elder and two-to-three family members who provided the elder with care but did not live with the elder or each other. Each participant was interviewed at the beginning and end of the three weeks and also completed some questionnaires. Four sets of families participated (i.e., four elders plus two-to-three family members per elder); the deployments were conducted from September to December 2003.

Multiple aspects of our study used Wizard-of-Oz techniques. First, when we were doing this work in 2003, digital picture frames were pretty basic and unable to handle what we had in mind. To simulate an interactive digital picture frame, our prototypes were built using a touch-screen tablet PC housed in a custom-built beech wood picture frame. The display itself was shown via a web browser running in full-screen mode so that any distinguishing browser characteristics were hidden; the tablet couldn't be used for anything else unless it was removed from the wooden frame.[17] We equipped each prototype with its own wireless GPRS (General Packet Radio Service) card for always-on internet access so the display could be updated throughout the day without needing to use participants' phone lines or requiring participants to have broadband internet access (which was not common at the time). Basically, the prototype looked and acted like an interactive digital picture frame even though such a frame didn't yet exist.

---

[16] The types of information about the elder that were shown in the CareNet Display came from results of a card sorting exercise in the roundtable discussions.

[17] Removing the tablet from the wooden frame required a philips head screwdriver and breaking through a label that asked the participant not to remove the tablet from the frame, but rather to get in touch with the researchers instead (and the label included our contact information). In spite of these deterrents, some participants removed the tablet without contacting us first.

Second, since the sensing and inference wasn't ready yet, to collect the data that was shown on the displays, a researcher spoke to the elder by phone three to six times per day everyday of the three week deployments, including weekends and holidays, and then immediately updated the displays after each phone call. To make sure that we were updating the displays with realistic data, we worked with the researchers at our lab who were developing the sensing and inference for the project to ensure that the type and level of detail we were collecting would be reasonable for sensors in the near future to provide.

Our approach required a pretty substantial effort both from the researchers and the elders who participated. However, it was MUCH less than the effort that would have been required to build a fully functioning prototype, deploy it, and then learn some of the same things that we learned from our Wizard-of-Oz deployment. For example, one of our key lessons was that one of the types of information we provided in the display wasn't very helpful, at least not in the display. In the card sorting exercise from our roundtable discussions, *falls* were the #1 type of information family members wanted to know about the elders. However, even frequent falls aren't that frequent (e.g., when compared to daily meals or medications), and participants thought other information (e.g., household chores that the elder needed someone to perform) would have been more helpful. We also found that participants who kept their CareNet Display in their TV room or within view of their bedroom were pretty annoyed by how bright—and therefore distracting—the CareNet Display's screen was. In fact, the bright screen in a dark room was the number one complaint about the CareNet Display from the field study (see Figure 3.5).

Figure 3.5: The CareNet Display prototype in a dark room. Participants #1 complaint about the CareNet Display was how its bright screen unnecessarily drew their attention in a dark room (e.g., while trying to watch a movie or go to sleep). From Consolvo and Towle (2005).

To see how easy it would have been for experts to predict the usability issues found during our Wizard-of-Oz deployments, we compared our field study's findings to the findings of a heuristic evaluation of the CareNet Display that was performed by eight expert evaluators using Mankoff et al.'s *Heuristic Evaluation of Ambient Displays* (2003). Many of the usability problems were predicted by the expert evaluators—including the issue with falls—but the single biggest complaint that participants voiced in the field deployment—the bright screen in a dark room—was not predicted by anyone (Consolvo and Towle, 2005).

In spite of the effort that was required, we learned a huge amount from our Wizard-of-Oz deployments that informed much of our subsequent work in ubiquitous computing including our lab's *Technology and Long-term Care* (TLC) project (Reder et al., 2010).

3.4   LAB USABILITY STUDIES

Throughout this book, it is probably obvious that the authors are strong advocates of conducting mobile user research in the field—from early exploratory studies to understand the context for which you're going to be designing to evaluative studies to determine how well or poorly a technology works (or might work) out in the wild. However, field studies are often expensive, time consuming, and a lot of effort. If you have a technology to evaluate—especially something new or a new take on something established—it's often advisable to conduct a lab usability study(ies) *before* going into an evaluative field study so that you don't waste precious resources uncovering issues in the field that could have been uncovered in the lab. This way, you can spend your evaluative field study learning what you couldn't in the lab.

Perhaps the most common lab-based user research method in HCI is the lab-based usability study. Usability studies (or usability testing) have been around for decades and are commonly used for evaluating desktop software and websites—and more recently, mobile applications (e.g., Thompson et al, 2013). Rubin (1994) defines usability testing as a "process that employs participants who are representative of the target population to evaluate the degree to which a product meets specific usability criteria."

The gist of a typical lab-based usability study is that the researcher recruits people who represent the target user population,[18] then each participant comes to the lab for a one-on-one session with the researcher (sessions are usually about 60 min long; more than 90 min, and you seriously risk fatiguing the participant). The session begins with the participant getting the time to read, ask questions about, then sign a consent form (and possibly an NDA or other paperwork, depending on the study and the study's sponsor). Some researchers like to send the consent form and any other paperwork in advance of the one-on-one session and ask the participant to sign before they arrive

---

[18]   8–16 participants is common for a single study, however, the exact number of participants you need depends on your research questions, the design of your study, and what you're trying to learn.

to save time and to ensure that the participant is okay with what they're being asked to sign. The researcher then asks the participant to do representative tasks with whatever technology is being evaluated. While the participant is doing these tasks, the researcher should silently observe, keeping a careful watch for what the participant does, where they succeed, and where they fail. It's not uncommon to ask participants to talk through what they're thinking (or "think aloud" (Nielsen, 1993; Nielsen, 2012)) as they try to complete the tasks. In most cases, the researcher should not intervene if the participant struggles or has a question. If the researcher does intervene, it's advisable to note that the researcher had to intervene and consider any results after that intervention appropriately. After the task portion of the session is over, the researcher may wish to conduct a short interview with the participant and/or ask them to complete a short survey about their experiences.

Usability studies can be conducted on paper sketches of a proposed interface (Rettig, 1994) using the Wizard of Oz technique discussed above, working deployments of existing applications or systems, or anything in between. The goal of this chapter is not to teach you how to conduct a usability study, as plenty of great resources for that already exist (see Chapter 1 for resource recommendations about how to conduct usability studies and tips for recruiting participants as well as obtaining informed consent). So instead of giving the reader a lesson in how to run a lab-based usability study, we will instead point out some limitations and offer some tips that are particularly relevant to lab-based usability studies of mobile technologies, and provide a case study that used a lab usability study to prepare a system for a subsequent field study.

3.4.1 LIMITATIONS

While an in-lab usability study can get you quick feedback from a variety of people, and generally can find the major usability issues with fewer than a dozen participants, the method has clear limitations, particularly for mobile user research. Because participants are in a lab setting, any environmental usability issues will likely not be found. For example, imagine a location-based content system where there just is no content in the places where a person wants to use the system. Or that there is poor network connectivity in these places, restricting the use of the system if it's not pre-caching content.

Social acceptability is also difficult to test in the lab. Imagine a system that primarily uses voice to navigate to particular features, similar to some smartwatches that have recently come on the market. Perhaps the participant can easily figure out the voice commands and execute them flawlessly in the lab, but the idea of talking to a phone or watch in public is just something they would never do. The often silent conditions of a lab might also be very different from noisy environments where the system might be used in actual practice. Other conditions of the real world are difficult to properly simulate in the lab. For example, bright glare from the sun on a screen or road vibrations while driving or on a train.

Lab studies are also generally one-time affairs. Participants will see your system for the first time and, as researchers, you can understand what's known as "initial usability." However, many applications have some learning curve and the third or fourth time a person interacts with a system might be a very different experience than the first. Since most users will not be first time users, better understanding the use that evolves over time is often a much better indicator to the future success and overall usability of the system.

In-lab usability testing can be a helpful and cost-effective step in developing a new system. But whenever possible, it should be augmented with evaluative field study methods that were described above to better understand usability issues that only surface in certain contexts or complex environmental and social aspects of use that might be even more serious than a usability issue.

3.4.2   LAB USABILITY STUDY TIPS

After your exploratory fieldwork has helped you develop a good idea of what you want to design, we advocate starting early and iterating. As we mentioned, usability studies can be conducted on a broad range of system fidelities—even for mobile technologies (the case study below is about a lab usability study that we conducted with a mostly paper prototype). Getting feedback early—even on sketches—then iterating will help set your project up for success.

Two of the most common mobile technologies tested in lab usability studies now are mobile phones and tablets. When you're going to conduct a usability study with either, one of the first decisions you have to make is whether the participant will perform the usability tasks on their device or if you're going to provide them with a device. The right decision will inevitably depend on what you're trying to evaluate. If you provide them with a device, we recommend keeping electronics-safe cleaning wipes nearby and wiping the device down after each participant. It's a good idea to wipe the device down right before you hand it to them, so that they can see that you've cleaned it, then do it again after they're done, so they don't need to worry about smudge attacks (Aviv et al., 2010), especially if they've had to authenticate (e.g., by providing their fingerprint or typing in a passcode) at some point during the study.

If you do use a lab-provided device, try to recruit participants who have the same device model, home screen launcher, and operating system version to limit effects of unfamiliarity with the device itself and focus on the application that you're testing. In general, it's not recommended to have iOS users try to use an Android prototype or vice versa given the differences in interactions and device size.

Whether they use your device or their own, you'll also need to decide if you want to video record what they do on the device—or simply stream an image of what they're doing to another screen so that you can watch. If you do want to video record their screen, a common tool to use is a mobile sled (see Figure 3.6); there are plenty of tutorials online to walk you through making your own sled based on the size that you need. It's a good idea to place some reusable adhesive dots (e.g., GekkoDots[19]) on the base of the sled to help keep the device in place without damaging the phone. One final tip about video streaming or recording is to be careful not to capture the participant authenticating anywhere with their own account credentials or device security code (remember that onscreen keyboards briefly reveal each key typed, such that a password could be retrieved from video). When it's time for them to authenticate, block the camera, have them remove the device from the sled, unplug the cord that's supporting the screen streaming, etc. The same tip applies if they visit any potentially sensitive data (e.g., opening their email or social networking account). We frequently mention to participants that if any sensitive information appears that we did not block out in advance, to let us know and we will remove it from the video before archiving for other team members to view. This could happen in a study of email, for example, if an account number or potentially sensitive receipt shows up as part of using the product for a task.

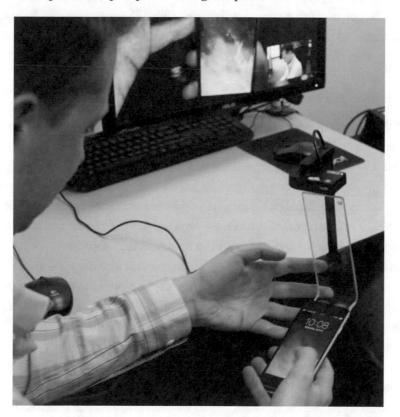

Figure 3.6: Example of a mobile sled. Mobile sleds allow the participant to pick up and interact with the phone in a more natural way, while video recording the screen and finger interactions. Courtesy of Yahoo, Inc.

[19]   www.gekkodots.com {link verified Nov 27, 2016}.

And as you will see below, even in the lab, it's possible to partially simulate more realistic environments if that's appropriate for your study. This can mean having participants perform tasks with a technology while walking on a treadmill if you intend for them to use the technology while walking out in the field, or perhaps sitting down on a sofa in the lab to help approximate what it might be like in their living room. Still, these types of experiments are far from the real world of walking in a crowded city, using a phone on a busy bus or subway, or other everyday scenarios, so interpret the results accordingly.

In general, try to simulate real world use as best as possible. Be careful of how you describe tasks to your participants so as not to bias them down a particular path in the application. For example, if the product will prompt them to log in with a button labeled "Sign In," use different language to describe the task. Or better still, ask them to "get to their email" or whatever product you're testing. To a typical person, that is the task that they are trying to perform, and signing in is just a road block to get there that they usually won't think about in advance. Focusing on the real task the person wants to perform will lead to more natural reactions when things intervene along the way to getting their goals accomplished.

Example: Home Energy Tutor Wizard of Oz Lab Study

In Summer 2003, we were developing a tool at Intel Labs Seattle—the *Home Energy Tutor*—to help users monitor, understand, and ideally reduce their energy use at home (Beckmann et al., 2004). The concept was that a homeowner would be shipped a Home Energy Tutor kit, which would include sensors, instructions, and supporting computing infrastructure. The homeowner would install the sensors and run the system for about a month, at which point they would pack it up and return the kit to the energy utility or other sponsoring organization. Set-up of the kit would require the homeowner to place sensors on appliances and in rooms around their home. As part of the set up, they would also have to create associations between each sensor and the appliance it was on or room it was in. Since this was envisioned to be a temporary deployment where the kit would be returned for others to use, it needed to be safe and easy to set up and remove, and it couldn't require access to hidden outlets or electrical cords, the electrical mains or breaker box, or other places that would be difficult to reach.

Our idea for how the association would happen during set up was that the user would walk through a wizard on a personal digital assistant[20] (PDA) that would instruct them to scan barcodes on the sensor and in a printed catalog with the PDA's attached barcode scanner. The sensor installation task, in particular, our idea about how to associate the sensor with the appliance it was on or

---

[20]   A personal digital assistant (PDA) was a handheld computing device that was primarily used for personal information management. Some had Internet connectivity and/or could be synced with a laptop/desktop. They were quite popular in the mid-1990's to mid-2000's; their popularity waned as mobile phones became more feature-rich and widely adopted. The photos in Figures 3.7e and 3.7f include a PDA.

room it was in, was the focus of our first study—a lab-based usability study (Beckmann and Consolvo, 2003). To conduct our evaluation, we created low-fidelity prototypes of many components of the system, as follows (see Figures 3.7a, b, and c):

- *quick-start guide:* a single sheet of paper with a high-level summary of the instructions;

- *an item catalog:* a paper-based set of pictographic representations and text descriptions of appliances and locations in a typical home;

- *sensors:* two low-fidelity mock-ups each of the vibration, current, and motion sensors that we intended the system to use. The "current" sensor mock-ups were simply grounded-plug adapters that we painted yellow. The "vibration" and "motion" sensor mock-ups were made of particle board to represent the rough size and shape that we expected the final sensors to be; and

- *handheld scanner with screens:* a particle board mockup of the PDA + attached barcode scanner that we intended to use in the system, as well as paper mock-ups of the wizard screens that the handheld scanner would display.

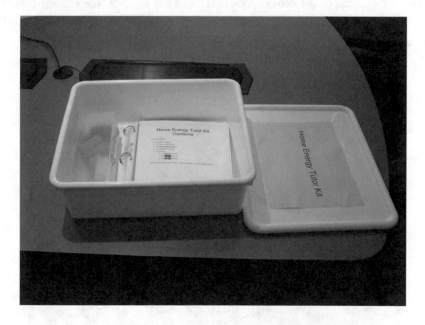

Figure 3.7a: Low-fi version of the Home Energy Tutor sensor installation kit. Contents of the sensor installation kit, as seen by participants. From Beckmann and Consolvo (2003).

Figure 3.7b: Low-fi versions of the item catalog and handheld scanner. The scanner mockup is an out-line of the hardware we intended to use, traced onto then cut out of particle board. From Beckmann and Consolvo (2003).

Figure 3.7c: Low-fi versions of the sensors. From left: motion sensors (blue, with removable adhesive backing), vibration sensors (red, with magnetic backing), and current sensors (yellow). From Beck-mann and Consolvo (2003).

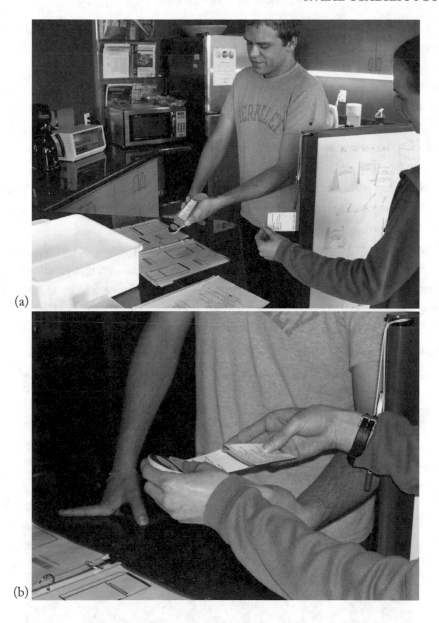

Figure 3.7d: The Wizard-of-Oz technique in action. (a) The participant (at left) uses the particleboard mockup of the handheld scanner to "scan" a barcode from the item catalog, while the "computer" (at right) prepares to change the screen in response to the participant's action. The rolling whiteboard in front of the "computer" contains the various screens that the computer might need to use depending on what the participant does. (b) The "computer" changes a screen on the handheld scanner in response to the participant's prior action. The moderator, who also served as the note-taker, is nearby but out of view. From Beckmann and Consolvo (2003).

We started with two pilot sessions (one with a co-worker and the other with a family member of a co-worker), and revised the protocol for the study accordingly, which included abandoning our plan to video record sessions in response to concerns mentioned by our pilot participants. For the study itself, eight homeowners participated. Each session, which took place one participant at a time in our lab, consisted of five parts: (1) consent paperwork; (2) a questionnaire; (3) three Wizard of Oz-style (WOz) task-directed sensor installation exercises (see Figure 3.7d); (4) another task-directed exercise to evaluate the usability of the actual PDA + barcode scanner that we intended to use (see Figure 3.7f); and (5) a post-task interview about their experiences. The WOz tasks were attempted in our lab's kitchenette (which included a refrigerator, sink, toaster oven, and microwave oven) and quiet room (which included a sofa, end table, and lamp) to mimic aspects of the home environment.

As a result of our lab usability study, we made many improvements to the interface on the handheld scanner and to our paper documentation, as well as changing the adhesive on the sensors in response to concerns raised by participants. We built a higher-fidelity version of the Home Energy Tutor installation kit (see Figures 3.7e and 3.7f), which we evaluated in the homes of 15 Seattle-area homeowners (Beckmann et al., 2004). Our lab usability study enabled us to fix basic usability issues with our system before going into the field.

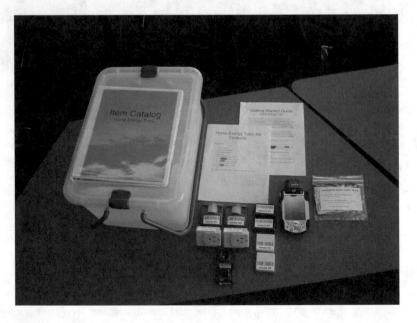

Figure 3.7e: Higher-fidelity version of the Home Energy Tutor sensor installation kit. Following the WOz lab usability study, we developed a higher fidelity prototype of the Home Energy Tutor system that we tested in a WOz field-based usability study. From Beckmann et al. (2004).

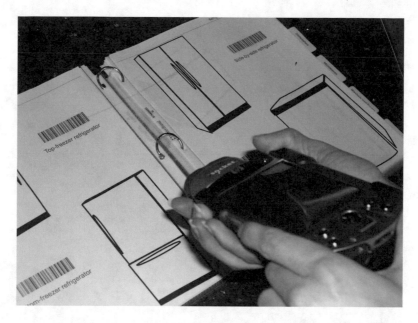

Figure 3.7f: Scanning an item with the Home Energy Tutor's higher fidelity prototype. The higher-fidelity version of the task shown in Figure 3.7d above—with a working PDA and barcode scanner. From Beckmann et al. (2004).

3.5   SUMMARY

In this chapter, we discussed *exploratory* or *generative field studies* that help you understand the people and environment you're going to design for, including where and how they spend their time and the contexts in which they're likely to use your technology. Exploratory field studies are great to conduct at the early stages of a project, before you fully develop your ideas about what you want to design. Such studies can help you generate great ideas that you might not otherwise think of (e.g., they help you go beyond your own experiences and those of your research team).

We also discussed *evaluative field studies*, which investigate how people engage with a technology—perhaps the technology that you designed—in the wild, including how it fits into their lives. And finally, we discussed *lab usability studies*, which we encourage you to conduct before going into the field for an evaluative study if you have something new or a new take on something established to evaluate. Although field studies are very important for mobile research, they are expensive and time consuming—you don't want to waste precious resources uncovering issues in the field that could have been uncovered in the lab. This way, you can spend your precious field study resources uncovering what the lab won't.

CHAPTER 4

Diary Studies and Experience Sampling

4.1 INTRODUCTION

In this chapter, we describe two self-report user study methods that can be used in the field to understand behavior over longer time periods (typically one week or more, and sometimes one month or more). We begin with *diary studies*, where participants keep a log (text, voice notes, photos, etc.) about their thoughts or experiences (e.g., the times and places that they exhibited a certain behavior). We then discuss the *experience sampling method* (ESM), where participants respond to brief questionnaires at specific times or contexts. These two methods share many things in common, and they can be used to collect similar types of data (including qualitative and/or quantitative).

As mentioned in earlier chapters, diary studies and experience sampling form the backbone of many field studies in the Human-Computer Interaction (HCI) domain. They allow researchers to hear directly from the participants in a study while the study is in progress and reduce the reliance on recall from other common self-report methods, such as traditional interviews and surveys. They help researchers to more reliably understand how and why a participant performs certain actions at specific times under naturalistic circumstances as well as how attitudes or behaviors change over time in a way that a single interview, survey, or quantitative logs cannot fully address.

4.2 DIARY STUDIES

Diary studies are one way to learn about what people are doing out in the world, close to the time those interactions naturally occur. In a diary study, researchers ask participants to keep a log of their attitudes or behaviors related to a topic area of interest, often for several weeks, to help the researcher understand more about that topic. Diaries can also be helpful for capturing emotional reactions to specific interactions close to when they happen. Simply asking the participant to recall something about use several weeks later in an interview or survey often will miss key details of specific interactions and can result in untrustworthy estimates of how often a person performs a specific task or how they felt at the time. Diary studies can be used both in early, generative stages of research to understand current thoughts and practices (e.g., around phone messaging use, photo sharing, etc.) or in later evaluative stages when seeking to understand how people use a particular technology in their daily lives.

The concept of a diary study long predates the mobile phone. In the early 1900's, several diary-style studies emerged to explore how people were spending their time. One of the earliest was a study in the UK (Reeves, 1979) that explored how women with limited financial means were spending their time (and money). Thirty-nine women were asked to complete a daily diary of activities and expenditures that researchers used to better understand the patterns of their days. This method has been repeated throughout the world in a variety of *time use studies* that have explored how daily life has changed over the decades. Similar to the work of Reeves, these time use studies ask participants to categorize each chunk of their day into one of a set of predefined activities such as working, sleeping, watching television, preparing meals, and so on. A fairly comprehensive list of these studies can be found in Fisher and Tucker (2013).

These diary methods were brought into the field of HCI by Don Norman (1981), Abigail Sellen (1994), and others in the 1980's and 90's. This early work was largely composed of studies about the workplace and understanding the timing, frequency, and type of tasks workers faced throughout the day. Participants were usually asked to keep a paper diary and to create an entry whenever they performed a specific task that was of interest to the researchers for a specific study. These diary entries helped to explore the context of a particular task and the steps that the research participants took to address the task. This data was then used to create better work support tools that fit into the structure of the participants' work lives.

Diary studies entered the world of mobile user research through the work of Palen and Salzman (2002). The researchers used voicemail-based diaries, as they offer advantages over carrying around paper and pen. Mobile phones make calling a simple, almost-always available, task. Since mobile interactions, much like workplace interactions, occur over longer periods of time and cannot easily be directly observed by researchers, these methods offer researchers a lens into the daily lives of participants, to help them understand how people interact with technology and with each other. Much of our use of voicemail diaries in mobile situations was influenced by this work.

Diary studies can be helpful, and they often require less of a daily burden for participants over experience sampling (described later in this chapter), when in-the-moment data collection is not needed and a summary at the end of the day will suffice. Diaries can help participants to reflect on their behaviors, close to the time of interaction, but can occur when they have free time. This allows participants to live their lives throughout the day and not be interrupted in the middle of a task by questions from researchers. This can be useful when the specific questions that they should address in the diary are not context specific, but rather ask them to reflect on their everyday interactions throughout a day or week. It can also allow a wider variety of participants to take part in your study, as those who frequently meet with clients, are in the medical field, or otherwise uninterruptable throughout the day can also take part.

4.2.1   BASIC METHOD

Diaries can take many forms, with the overall goal of understanding the experience that a person is having out in the world at a given time. These days, the most common diaries are voice-based diaries, implemented either as voice notes or through a voicemail number. Other types of diaries can include paper-based logs, photos, or videos captured at the point of interaction, photos of communication artifacts such as letters and cards, creative tasks that participants are asked to perform throughout a study, or structured survey questions asked daily, if there is desire for more quantitative information. Most commonly, diaries are combined with other methods such as interviews and instrumented applications to help fill in a deeper qualitative understanding of specific instances of use that goes beyond what can be understood from usage logs or recalled in interviews alone.

Designing Questions

The first step in planning a study that will involve a diary component is to consider the role the diary will play in the larger study. Which research questions can be answered with diary data? How will diary data combine with the data collected from the other methods in the study? Thinking deeply about these questions can lead to the specific questions to ask during the study as well as the frequency with which you want to ask them.

For example, you might be interested in understanding more about people's calling patterns and why they choose to call certain people at certain times and places. A diary study that asks participants to bring up their recent call list each evening and discuss the context around each call could help to answer these questions. Waiting longer than a day risks the participants forgetting the specific details around a particular call, while asking people to call a voicemail after every phone call likely will involve prohibitive effort or even modify their calling patterns to avoid having to place a second call if they are in a rush.

In another study, you might be interested in understanding the qualities of the types of places that a person visits, perhaps to create better local recommendation systems. For this type of study, perhaps asking participants to review photos or video clips that they take throughout a week might help elicit specific visual or experiential aspects of the places that people enjoy visiting. Participants could be shown photos at the end of each day and asked to elaborate on why they chose particular places to visit or capture media. This could be combined with final interviews where participants expand on what they like about each place and can serve as reminders of the places that they visited, instead of asking participants to try to recall places that they visited.

In choosing questions, it is important to keep participants' daily obligations to a minimum. We have found that keeping voicemail diaries to about a minute or two per day serves to get the basics of an experience while not putting too much of a burden on participants. You can usually follow up with a participant to get more detail on interesting stories in an interview conducted when the diary portion of the study is over. Keep this in mind as you craft your questions. We have also found that *daily* voicemails, especially over two weeks or more, seem to be a maximum frequency for not becoming too burdensome. We recommend that you craft the questions that participants should answer in ways that encourage them to talk, rather than simply responding to yes/no questions. Asking "why," "how," or other rich and open-ended questions will encourage people to explain in more detail about what they have recently experienced. The researchers should also state that it is perfectly acceptable for participants to report that they did not do anything that you asked them about in a particular day, and that they won't experience any penalty in their participation incentive for doing so. Giving this as an explicit option helps reduce the chance of participants doing something unnatural or making something up to report on so that they feel that they are complying with the study to get their full incentive.

Collecting Data

We have found that a system where diary entries are immediately available to researchers can help to ensure that a study is progressing and that participants are not experiencing any problems with the data collection. Generally, we set up a phone number with voicemail for participants to call in to leave their messages. Depending on the service you use, you can often set it up to generate an email to the researchers when a new message has been left that includes an audio recording and crude transcript of the message. Many such services are available, and it's best to check with your institution on which tools and services you may use to collect data from participants.

Receiving an immediate transcript or audio file helps the research team, especially if the participant mentions any issues with an aspect of the study (e.g., the prototype application isn't working). This enables the researchers to react quickly and gives them the opportunity to fix a problem before it's too late. Also, receiving the data as the study goes along can help researchers to prepare for final interviews, add additional questions to the final interview protocol for unexpected themes that arise, and can lessen the last minute scramble before a final interview. It can also help the researchers prepare the participant's incentive, particularly if the incentive is based on compliance to diary frequency.

However, without a specific reminder, participants often forget to call in to the voicemail system. Before smartphones with background applications and reminders became common, we often gave participants a printed piece of paper to place on their coffee table or nightstand to remind them to call at the end of the day. Now, however, it tends to be much easier just to send an email, a text message, or add a calendar appointment each evening (or at whatever interval you choose) with the phone number to call in the "Location" field and the questions that they should answer in the body of the calendar appointment (see Figure 4.1). During the initial interview, we generally let the participants choose what time they would like the reminder to appear to best fit with their schedule. Do not assume that the time that works best for you will work best for others, as many people work night shifts, go to bed much earlier or later on a regular basis, or have other obligations or routines that don't match yours.

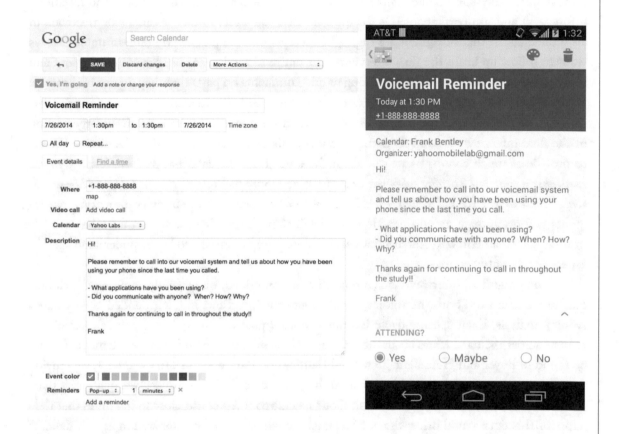

Figure 4.1: Setting up calendar reminders to call in to the voicemail diary. Left: Setting up a recurring calendar appointment for 14 days with a reminder to send to participants. Placing the phone number of the voicemail system in the "Where" field allows it to be clicked to dial when the reminder appears on the phone. The description summarizes what we'd like them to talk about. Right: The voicemail reminder on the phone. Participants can just click the phone number to call in, and the tasks for them to report on are summarized in the message body. Courtesy of Yahoo, Inc.

Preparing for Final Interviews

Once the study is underway, researchers should regularly review the data that is coming in. Generally, we try to transcribe voicemail diaries on a daily basis to keep up with our participants. Often, we use third party transcription services, such as rev.com,[21] especially when we have many participants running at the same time.[22] We keep all voicemail entries for the same participant in a single Word document or Google Doc, coded with the date and time of the voicemail. It is important to remove personally identifiable information (phone numbers, addresses, names of friends/relatives/businesses) and use participant IDs for the data that you store, so that it cannot be tied back to an individual. In that document, we highlight excerpts that are particularly interesting or that we want to follow up on in the final interviews. The entire research team can see each document and highlight aspects that they find interesting or add comments on parts that they do not understand.

When the final interview arrives, we bring these coded voicemail transcripts (or printouts from photo diaries) into the interviews with us, and we frequently have a relatively large section of the final interview reserved for asking about specific incidents mentioned in the diaries. We try to provide as much context as we can about the event from the data that we have (e.g., *"Can you remember a time last Tuesday when you called your mom from the train station in Oakland because she wasn't there to pick you up? Could you tell us more about this event?"*). In an interview setting, participants will often open up and provide more detail about a specific experience than what was left in their voicemail entry. Researchers can also ask follow-up questions to better understand specific aspects of the interaction.

Once the final interviews are completed and transcribed, we generally combine all relevant qualitative data from both the voicemail entries and the interviews into a grounded-theory based affinity analysis. Each specific quote is printed onto a post-it note and we iteratively search for themes across the data. More about the affinity analysis process and how we've implemented it can be found in Beyer and Holtzblatt (1998) and Bentley and Barrett (2012). In Chapter 1, we provide recommendations for other resources that can help you learn about how to analyze qualitative data.

Overall, diaries help to gather data about specific contexts of use, close to the times that they happen. This data would otherwise not be remembered, or not be remembered in as rich detail if asked about only in a final interview. Specific information such as understanding the frequency that people perform certain tasks or the emotional reaction to specific interactions is well suited to this technique.

4.2.2   VARIATIONS

Some variations exist to the basic method. The most common is using forms of data collection other than voice or using diaries that display content captured throughout the day for a reflection period each evening. Photo or contextual diaries are common variations to the method, where at the end of each day, participants are asked to review photos that they have taken, or logs about

---

[21]   Link verified Dec 27, 2016

[22]   As mentioned above, check with your institution on which tools and services you may use for any handling of participant data.

their location or other context that were captured during the day to provide additional detail to the researchers while the memories are still fresh. These reflections can help researchers understand the context of use in greater depth than the raw logs or often-faulty memories some weeks later in a final interview. Another variation is more of a "daily diary" style whereby the focus is on gathering information about an entire day. While this style of querying re-establishes a greater possibility of recall bias (e.g., asking for information about how stressed a person is throughout the day will be influenced by their stress in that moment more so than their stress at the end of the day), with that limitation acknowledged, it can still provide some valuable information. For example, we used a daily diary style of surveying within the MILES physical activity app interventions to glean insights about general factors that were useful or not for fostering behavior change (King et al. 2013, 2016).

Below, we discuss three variations of diary studies: reviewing interactions, cultural probes, and quantified self reporting.

Reviewing Interactions

Some researchers have explored a hybrid method that combines the diary study approach with data that is collected in-the-moment about the events of interest to the researchers. For example, Intille et al. (2002) wanted to investigate user preferences for kitchen redesigns. Rather than inter-rupt the participant with an alert at the time of the event of interest (as you would in experience sampling—discussed later in this chapter), the researchers used a tool they developed that captured an audio-visual snippet of the activity. These audio-visual snippets were later reviewed by the participant at which point they provided their preferences; the snippets were used to trigger the participant's memory of the moment when the snippet was captured. Moira McGregor and colleagues from the Mobile Life Centre in Stockholm (McGregor et al., 2014) used a similar approach in one of their studies. Instead of just asking users to take a video or a voice-note after an interaction, they instrumented individual phones to capture video of the screen as well as ambient audio while a phone was in use. This method enabled the researchers to understand the context of use out in the world far deeper than logs or traditional diaries could. Researchers could hear participants talking to friends or family while using the device or hear background audio of buses, music, or other en-vironmental sounds to better understand the participant's context and what else they were doing while using the application of interest. Participants were then asked to complete a diary each day where they reviewed each clip and provided additional detail about their location, who else was around them, and a bit more about the situation. At this time, they could also delete items from the log that they did not wish to share with the research team.

Obviously, this variation on the method requires the participants to share a lot more about their life and possibly the lives of others with the research team, and it therefore might not be ap-propriate for your study. The additional data that's captured in the moment of interest can help the researchers learn more about the interactions of interest *and* it can help improve participants' recall when it comes time to compose a diary entry. When using variations such as this, it is extremely important to allow participants the chance to remove potentially sensitive data *before* sharing with the research team and to check with your institution about if and how you need to obtain consent from other parties whose data might be captured (and similarly, if you need to provide those other parties with the ability to delete data that's captured about them).

Cultural Probes

Another variation of the diary study method is best expressed in the work of Bill Gaver and colleagues in their work on Cultural Probes (Gaver et al., 1999, 2004). This method often involves sending participants a "kit" of creative objects for them to use to explore an area of interest, often over several weeks. This kit might include a camera, postcards, maps, or other creative objects. Participants might be asked to mark on the map the places where they spend time, take photos of specific experiences that they have out in the world or at home, or other creative tasks over a particular duration of time. Probes such as maps and photos can gather similar data to a traditional diary study and can be fun for participants as well, as they get to use particular objects that have been given to them as a part of the study. These techniques often work best for early stage, exploratory research in understanding current practices and communities or testing out new technology concepts. We used a diary and cultural probe-like approach in some research we did at Intel Labs Seattle and the University of Washington to explore how people might react to potentially invasive sensors in their homes (Choe et al., 2012).

Quantified Self-Reporting

Although most of the above examples have focused on capturing qualitative data (e.g. how people feel, where or why they performed an action), other types of diary studies can ask participants to quantify specific aspects of their life on a regular basis. While it's generally better to try to record this data automatically where possible (e.g., people are generally not good at estimating their time spent on certain activities; see, for example, Klasnja et al., 2008), these types of diaries can help discover otherwise unnoticed trends or correlations among broader participant bases. For example, a diary study could ask someone to estimate the time that they spent driving that day, or to enter specific details that they could look up on their device, such as the number of emails that they sent that day.

4.2.3   LIMITATIONS

While diaries are a great method to learn from your participants between interviews, there are some limitations to the method that are important to consider. The largest limitation is that the act of keeping a diary might alter participants' behavior from what they would otherwise do. Participants often feel like they must have something to say each day, meaning that they might use your application or make an extra call just so they have something to talk about. In general, we try to mitigate this risk by setting the frequency of voicemail calling to be approximately close to expected application use. If we were fielding a fantasy football application, for example, we'd likely only have participants call once a week. It is also important to tell participants that it's okay to call in and say that they didn't use the system or don't have anything to report during the period since the last voicemail, and they shouldn't feel obliged to use your system just because they are in a study or because they think it would be helpful.

While diaries can serve to elicit many details of use throughout the day, participants can easily forget specific interactions, especially if they see them as mundane. For example, they might forget to report a phone call they made to coordinate dinner or pick up a child, as they do these things nearly everyday, and often at hectic times of the day. To help mitigate this, you might consider augmenting the diary with additional data for them to review when making an entry, as described above. Similarly, sometimes, despite leaving some information in a voicemail diary, weeks or months later during a final interview, participants often cannot recall the actual incident, even after reviewing their relevant diary entry. This is often unfortunate, as something you might find interesting cannot be followed up on in any additional detail. Often, the diary entry alone is not enough to fully understand the situation that occurred. Depending on what you're trying to learn and how well participants can recall the details you need them to recall, experience sampling (discussed below) might be a more appropriate method to use.

A related limitation is that voicemail diaries are often very short and can lack important details you might need to inspire your design or better understand what you're trying to study. Diaries almost always need to be combined with other methods of data collection, such as in-depth interviews *after* the diary portion of a study is over, to explore what was reported in the diaries in more detail.

4.2.4   CASE STUDIES

We often use diaries in our field studies. In this section, we describe how we used voicemail diaries in two specific studies. The first is an early-stage, generative study conducted at Yahoo Labs investigating how teens use their mobile phones. The second is a field study of the Serendipitous Family Stories application at Motorola Labs, designed to explore and understand the use of the system in everyday life.

Teen Study

In the Teen Study (Bentley et al., 2015) conducted in summer 2013, we were interested in details around all aspects of phone use and especially in understanding how phone use fit into teens' highly structured days as high school students. The study was structured with an initial interview, 14 days of a voicemail diary complemented with a logger on the participants' phones that collected details about the times and durations of application usage, and a final interview. Fourteen diverse teens from throughout the San Francisco Bay Area participated. We asked each participant to call in to our voicemail diary twice daily for two weeks to discuss how they were using their phones. The voicemail diary data complemented qualitative data from initial and final interviews as well as quantitative data from logging applications that we installed on their phones to collect data about which applications were used throughout the day, at what time, and for how long.

To remind them to call in to the voicemail diary, we created calendar reminders for our participants that they received twice a day at times of their choosing. We suggested that one could be when they left school for the day and another before they went to bed. Afternoon reminders ranged from 2:30pm to 6pm and night reminders ranged from 8pm to 11pm. These calendar reminders included the questions that we wanted them to answer: "*How did you use your phone since the last voicemail entry? What applications did you use? How did you communicate with others? Please tell us details of your interactions.*"

Over 14 days, we collected voicemail entries from each participant, transcribed the entries and highlighted points for follow-up as described above, and prepared for final interviews where we asked participants for more information about specific instances mentioned in the voicemails.

Through the voicemail diaries, we were able to learn much more about the day-to-day use of mobile phones than we could have from the initial and final interviews alone. We learned how mobile phones fit into everyday activities and allowed people to share moments with others. For example, participant 12 (or "P12") recalled how she posted a collage on Instagram of her softball team after practice:

> "*We all got really dirty at that practice, we were sliding and everything, so we all took pictures and then posted and made a collage.*"

P12 discussed liking the ability to be artistic on Instagram. She posted a photo from her grandmother's house:

> "*My grandma has a really cool bathroom. Like she's got a ton of mirrors ... you know those mirrors when it looks like a tunnel of you? That's what it looked like and I just felt like the art scene and just took a picture and then like ... I posted a picture on Instagram and saying a quote from [a Justin Timberlake] song. ... It's hard to explain. My artsy moment.*"

This quote led to great follow-up discussions in the final interview to get more detail of this specific interaction and more general patterns around being "artsy" using mobile photos.

Other participants helped us to better understand the content of photos that might not be remembered in a final interview. For example, P11 spoke in the voicemail diary about her messy room while preparing to leave for college:

> *"Yesterday I was packing and I knew it was a mess and I just took a picture of it and I was like "Packing for [city]" and I just sent it to everybody."*

In the final interviews, we asked follow-up questions about the reaction from her friends upon receiving this photo over Snapchat, enabling us to better understand how Snapchat led to conversations around photographs.

Sometimes, diary studies can uncover unexpected behaviors. For example, P13 took photos of his textbook before leaving school:

> *"I didn't want to bring the whole book with me, but if I just take a pictures of the pages then I can have it on my phone so I don't have to bring this textbook."*

This led to discussions in the final interview about other things he takes pictures of to avoid carrying books or papers around with him.

Overall, the voicemail diaries added a key component to our data collection. They helped us collect stories of everyday mobile phone use, fleeting instants that participants would likely not remember or think important if asked to simply remember the last few times they used their phones. Details from these voicemails were then used in final interviews to follow up on the points mentioned and to explore related topic areas such as the social communication enabled by photo sharing, artistic practices with mobile photography, or just taking a photo of a textbook page before leaving school.

Serendipitous Family Stories

In the Serendipitous Family Stories project at Motorola (Bentley et al., 2011), we created a system in 2010 that allowed parents or grandparents to share location-based video stories with their children or grandchildren. Through the system, they were able to record short videos discussing the importance of specific places in their lives that would be available to select recipients in the physical locations where the events occurred. Those recipients—i.e., their children or grandchildren—used a mobile application that would vibrate their phones whenever they approached one of these locations, enabling them to watch the video right in the spot where it took place. They could discover places such as the spot where their grandparents met, the location of their wedding, old workplaces, or the theater where they saw Frank Sinatra perform live, and they could hear about these stories while literally standing in the place of great importance to their older relatives.

We wanted to understand the feelings brought about by discovering a story from a parent or grandparent while walking about the city. While it would have been impossible to follow a person around and see particular chance encounters with stories over a one-month period, asking participants to call in immediately after finding a story helped us to understand their emotional reaction to the event as well as any follow up that they took. This study consisted of an initial interview with the grandparents or parents, where they recorded stories for the younger participant. We then met with the younger participant, installed the application on their phones, and had them use it for a month. During this time, we asked them to call in to our voicemail system on any day where they discovered a new story. Since we captured on our server when stories were uncovered, we were able to pay an additional incentive to participants if they called for the majority of stories that they discovered. We held final interviews with each participant at the end of the study. Ten younger participants who lived in the city of Chicago and their ten parents/grandparents who grew up in Chicago but now lived in Southern Florida took part in the study.

In this study, we programmed our voicemail number into each participant's phone as the alphabetically first entry for easy access. During the initial interview, we gave each participant a half-page description of what we wanted to know along with the voicemail number and encouraged them to place it on their refrigerator or coffee table so that they would not forget to call at least sometime during the day when they found a story. However, we encouraged them to call us soon after the event of finding a new story out in the world so that they would remember exactly what happened. Particularly, we asked participants to *"Describe the experience of finding the story. Where were you? What was the story about? How did you react to finding the story?"* In addition, we asked the parents or grandparents to call us if their children or grandchildren ever talked to them about a story so that we could get a second view on conversations that might be inspired by finding a story.

Through these voicemails, we learned rich details about the process of finding a location-based story. We learned about what it was like to come across a story, such as P2's description:

> *"I was surprised. I was not intentionally going to the location to see the story but it was a surprise for me ... and then this thing (the story) was right there!"*

Other participants told us stories of being frustrated about not getting close enough to unlock a story during their morning commute and continually getting notifications that they were "close" to a story, but not close enough to watch it. These reports helped us to improve the notification functionality of the application before our public release.

P6 told us of a phone call she received from her daughter-in law. Through this voicemail, we were able to observe the rich emotions that finding a story could bring about and the communication that ensued:

"[My daughter-in-law] *called to say that I made her cry that she was so touched by the stories. And she thought they were funny, the first ones I told her about.*"

P10 reported a similar emotional reaction:

"*After that* [watching the two new stories] *I called both my mom and my brother and I laughed about the situation 'cause it was pretty funny what they said.*"

P4 learned a lot about her grandmother's life growing up.

"*I didn't know that they didn't have a car. She never had a car growing up. So everything they did they took a bus or the train and things like that.*"

Through these and dozens of other voicemails about finding a story, we learned about the participants' initial reactions to finding a story from their parents or grandparents out in the world. The rich details about in-the-moment emotional reactions were much more reliable than asking about details weeks later in a final interview and helped give us additional topics to discuss during the final interviews to gain even more detail about these specific experiences. Together, data from the voicemails and interviews were used to create StoryPlace.me (Bentley and Basapur, 2012), a publicly available version of this concept that was released several months after the study.

4.3   THE EXPERIENCE SAMPLING METHOD

The *experience sampling method* (ESM) is another self-report research method that can be used to learn about the thoughts, feelings, actions, and experiences that people have out in the world (Csikszentmihalyi and Larson, 1987; Barrett and Barrett, 2001). It has also been called *time sampling*, *beeper studies*, and *ecological momentary assessment* (*EMA)* (Stone and Shiffman, 1994). With experience sampling, participants complete brief questionnaires[23] as they go about their everyday lives, often in response to being alerted. These alerts can occur several times per day and may go on for several weeks depending on the needs of the study.

Although experience sampling is a type of self-report research method, it minimizes (and possibly even eliminates) standard limitations such as recall or observer bias that are often present in traditional self-report methods such as interviews, surveys, and even some diary studies.[24] With experience sampling, participants are usually asked about their *current* thoughts, feelings, actions, or experiences *in the context* in which these thoughts, feelings, actions, or experiences occur. That is, when the participant is providing a response to a questionnaire, there is little to no reliance on memory, there are no researchers present, and the environment is not artificial or otherwise out of

---

[23]   A good rule of thumb is that questionnaires should take no more than 2 min to complete, often less. Running pilot studies will help you determine a reasonable length for your particular study.

[24]   As suggested in the section above, diary studies typically suffer from these biases less than traditional interviews and surveys do.

context. And because participants often respond to the same questions over time and in various contexts, the research team can measure how consistent the participant's responses are. If responses for an individual participant vary, the team may be able to determine the circumstances under which responses vary.

Experience sampling has been used by Human-Computer Interaction (HCI) and Ubiquitous Computing (ubicomp) researchers to investigate a broad range of issues including how interruptible people are at work (Hudson et al., 2002, 2003), what electronic displays are around people throughout their day (Consolvo and Walker, 2003), how people would want to respond to requests for their location from people in their social network (Consolvo et al., 2005), people's privacy concerns around audio recording (Iachello et al., 2006), drivers' experiences during car trips (Meschtscherjakov et al., 2012, 2013), stress (Weppner et al., 2013), and many more.

In this section, we start with a brief history of the experience sampling method, describe the basic method and key decisions that the research team needs to make, discuss its main limitations, then provide some case studies of how we've used it.

4.3.1   HISTORY

Experience sampling was originally used to help researchers in the fields of psychology and behavioral science better understand areas such as mood, time use, and social interactions. In the early days of experience sampling, questionnaires were typically administered with paper-and-pencil where participants were asked to respond at pre-scheduled times throughout the day or in response to particular events of interest (Nezlek et al., 1983; Wheeler and Reis, 1991). As mobile computing started to mature and become more broadly available, researchers began to use electronic pagers (i.e., "beepers") (Csikszentmihalyi and Larson, 1987) or watches (de Vries et al., 1990) to deliver alerts to participants. In some cases, instead of responding to questionnaires with paper-and-pencil, participants were asked to call a designated phone number to complete their questionnaire (similar to the voicemail diary studies discussed above) or listen and possibly respond to the questionnaire via an audio recording device (e.g., a cassette player).

In the late 1990's and early 2000's, as handheld electronic devices were becoming more common and powerful, some researchers made a shift to *Computerized Experience Sampling* (Barrett and Barrett, 2001). With computerized experience sampling, handheld electronic devices—such as PDAs—were used to alert the participant that it was time to complete the questionnaire as well as actually administer the questionnaire and collect participant responses. In addition to being able to support precise delivery of randomly timed or pre-scheduled alerts, computerized experience sampling made it possible for researchers to know how long it took a participant to respond to an alert, complete a questionnaire, and the researcher could even set a timeout for the questionnaire in the case of an inadequate response time (e.g., if the participant doesn't respond within X minutes of the alert, the experience sampling software could cancel the questionnaire and record it as being missed). Even the earliest versions of computerized experience sampling, such as Barrett and Barrett's *Experience Sampling Program* (ESP) (2001), improved the researcher's ability to track compliance and reduced the potential for problems being introduced through data entry errors (e.g., from the researchers having to transcribe paper-and-pencil or audio responses for analysis).

Experience sampling, and variations of it, emerged as a popular method for HCI and ubicomp researchers in the early 2000's. For example, James Hudson et al. (2002) used experience sampling implemented on RIM Blackberry devices to investigate the availability and interruptibility of 12 managers at IBM Research over the course of one week. Specifically, the researchers were interested in managers' attitudes toward being interrupted in different situations as well as how they spend their time at work more generally. Participants were alerted 10 times per day at random intervals between the hours of 8am and 9pm via a silent, vibrating alarm on the Blackberry. The 8-question surveys were designed to be completed in about 30 sec. Participants were asked questions such as (Hudson et al., 2002, p. 99):

- Right now, I am:

  ○ by myself

  ○ engaged with 1 other person...

    • This is a planned event.

    • This is an unplanned event.

  ○ engaged with 2 or 3 others...

    • This is a planned event.

    • This is an unplanned event.

  ○ engaged with many others...

    • This is a planned event.

    • This is an unplanned event.

- This activity is:

  ○ business

  ○ personal

  ○ other

Scott Hudson et al. (2003) also used experience sampling to investigate interruptibility, but they were focused on people with high-level staff positions at Carnegie Mellon University. The researchers' primary goal was to see if they could use sensing and inference to develop models that could reasonably predict a person's interruptibility. The experience sampling portion of their study was implemented on a personal computer (PC) that the researchers provided for the study. Participants were asked a single question (Hudson et al., 2003, p. 259):

"Rate your current interruptibility on a scale from one to five, with one being most interruptible."

Participants could respond verbally or by holding up one to five fingers, depending on their current response. Participants were asked this question two times per hour on average. The PC sat on the participant's desk and collected data from audio and visual sensors in addition to responses to the experience sampling. Four people participated in the study, which ran for about 2–3 weeks per participant.

Our first study that used experience sampling was conducted in August and September of 2002 (Consolvo and Walker, 2003). We investigated people's information needs and what output devices were available to them throughout the day (in an effort to support Want et al.'s *Personal Server* (2002)). Our study consisted of a one hour-long pre-study interview, one week of experience sampling, and a one-hour exit interview. Participants received 10 randomly scheduled alerts each day over a 12-hour time window where they responded to questions such as (Consolvo and Walker, 2003, p. 28):

- Where are you?

- Which of the following are available right now?

  - printer
  - desktop computer
  - laptop
  - video projector
  - PDA (other than this one)
  - television

Each time window was divided into 72-min intervals, during which the participant would receive an alert. We used Palm m500s and a separate digital camera for the experience sampling (i.e., participants had to carry both the Palm m500 *and* the digital camera with them throughout each day for this study). The Palm m500s ran our *iESP* software to alert participants, deliver questionnaires, and capture responses. We also asked participants to use the digital camera to take a photo of what they were doing when they received an alert.

As mentioned earlier, the experience sampling method continues to be used by HCI and ubicomp researchers to investigate a broad range of issues. In many cases, the researchers develop their own custom experience-sampling tool for a particular study. However, another line of work that HCI and ubicomp researchers have pursued is to develop and make available tools to support experience sampling. For example, we extended Barrett and Barrett's (2001) ESP[25] for Palm Pilot and compatible PDAs to develop *iESP*,[26] which provided the researcher with more questionnaire options than the original ESP supported (e.g., iESP supported a broad range of question response types, branching, and the use of variables in questions and responses). We used iESP for the study we discussed earlier of people's information needs and what output devices were available to them throughout the day (Consolvo and Walker, 2003), as well as for a study of how people would want to respond to requests for their location (Consolvo et al., 2005), which we discuss below. Building on what we learned from iESP, Froehlich et al. (2007) developed *MyExperience*,[27] a tool for Windows Mobile 2005 devices that combined an experience sampling tool with the ability to passively log device usage, user context, and environmental sensor readings. Below, we describe how we used MyExperience in the development and evaluation of our UbiFit system (Consolvo et al., 2008a, 2008b, 2014). Carter et al. (2007) developed the *Momento* tool to support a broad range of ubicomp evaluations including experience sampling and diary studies. Fetter and Gross (2011) developed the *PRIMIExperience* system, which uses instant messaging clients to administer ESM. More recently, the Personal Analytics COmpanion (PACO) (Evans, 2016) is an application that runs on Android and iPhone devices. PACO supports different types of methods, including experience sampling.

4.3.2   BASIC METHOD

Like diary studies, experience sampling is often combined with other techniques such as pre-study interviews, post-study interviews, and/or logging participants' interactions with or behaviors sensed by a technology. In addition to determining what, if any, additional techniques will be used, you must make several logistical decisions about the actual experience sampling portion of your study. For example, you will need to decide how and when to *alert* participants that it's time to complete a questionnaire. You will also need to decide how to *deliver* the questionnaires to participants as well as how to *capture* participants' responses. These decisions should be based on the study's needs, the burden on participants, the constraints of the participants' environment, and the constraints of the tools being used. The very nature of experience sampling means that it interrupts people in their natural environments as they live their lives. It's important to keep that in mind when making these logistical decisions.

Alerting

When deciding how to alert the participant, you must choose the type of alert, when to alert, and how the alert will be delivered.

---

[25]   ESP is available online at http://www.experience-sampling.org/ {link verified 4 Dec 2016}.

[26]   Although iESP was originally made broadly available online, unfortunately, it is no longer available.

[27]   http://myexperience.sourceforge.net/ {link verified Dec 4, 2016}.

Types of alerts

Alerts can be time- or event-based. *Time-based alerts* can be random (e.g., 10 randomly spaced alerts each day) or scheduled (e.g., 10 evenly spaced alerts each day). *Event-based alerts* can be triggered by the system (e.g., the system triggers the alert when it detects that a particular event of interest has occurred) or triggered by the participant (e.g., the participant manually triggers the questionnaire after they detect that an event of interest has occurred). In general, scheduled and participant-triggered event-based alerts are more prone to bias than are random or system-triggered event-based alerts. With scheduled alerts (and in some cases, with system-triggered event-based alerts), participants can anticipate when the alert is coming and might modify their behavior or responses accordingly. With participant-triggered event-based alerts, the research team may not be able to confirm that the participant triggered the questionnaire when an event of interest actually occurred.

When to alert

You must decide the time period during which alerts should be delivered, if that time period might need to be different on different days or for different participants, and how many alerts to deliver per day. For example, if there is no compelling reason to wake the participant in the middle of the night, you should specify the times of day during which alerts can be delivered (e.g., 8am–9pm). It is often helpful to be able to specify different time periods for different days (e.g., 8am–9pm on weekdays and 9am–10pm on weekends) as well as for different participants (e.g., to accommodate participants who keep different schedules). Such flexibility is particularly important if the study is using time-based alerts. It's also a good idea to specify how many alerts will be delivered each day (e.g., no more than 10 alerts per day). When using event-based alerts, more sophisticated logic might be necessary, depending on how often the event of interest is likely to occur. And, finally, depending on the situation, you may need to specify a maximum number of alerts or dates over which alerts should occur for the entire study (e.g., stop alerting the participant after the 210th alert—i.e., 10 per day for 21 days). This can be helpful if you loan the participant equipment for the study that runs experience sampling software (e.g., if the participant cannot return the equipment as soon as they've completed what should have been their last alert; it can be annoying for the participant to continue to be alerted after that part of the study is supposed to be finished) or if they're using their own device and are not likely to uninstall the experience sampling software when the study is over (e.g., if they might use the same software to participate in other studies).

How to alert

You must also decide how to deliver alerts—a decision that should involve careful consideration of the environment(s) participants are likely to be in when they receive alerts. Common methods for delivering alerts include one or more of the following: audible alerts, tactile alerts, or alerts popping up or appearing on the device screen. As with alert schedules, it's often useful to have flexibility with this, and it may be a good idea to give the participant control over how they are alerted throughout the study. For example, if they are likely to be alerted while in a movie theatre or in a meeting at work, an alert that makes a noise may be inappropriate. However, if they are likely to keep the alerting device in their bag, a tactile alert may go unnoticed. Depending on the particular needs of the study, it may also be appropriate to provide participants with a way to temporarily snooze or dismiss alerts.

Delivering

When deciding how to deliver the questionnaires, you must consider several options including details about the questionnaire design and whether questionnaires should be read or listened to.

Questionnaire design details

Your needs regarding questionnaire design will help you choose how to deliver questionnaires. In particular, you need to make decisions about question and/or response option ordering, whether any contingencies or branching is needed, whether questions should be assigned a probability that they will be asked, and whether you need to be able to update or change questions during the study. For example, should the questions themselves be in fixed or random order? Are there any questions for which you want to randomize or randomly reverse the order in which response options appear? Should questions be asked based on how a participant responds to a prior question or on other questions that have been asked? Do any of the questions need to be assigned a probability that they will be asked? (For example, remember that each questionnaire should be able to be completed quickly; assigning probabilities to questions or using strategic branching can help manage the burden placed on the participant.) Might you need to be able to update the questionnaires during the study?

Written or audible (or both)?

You also need to decide whether the participant should read or listen to the questionnaires, or have the option to do either. For example, if the participant *listens* to the questions, is environmental noise likely to be a problem? Will the sound disrupt people nearby or draw unwanted attention to the participant? Will it take longer to listen to than to read the questions (and if so, is that a problem)? Are the questions sufficiently short and uncomplicated such that the participant can easily understand what's being asked without having to replay the question? If they have to call in to a phone number to hear the questions, is cost or service likely to be a problem?

Alternatively, if the participant is reading the questions, are there font size or contrast issues? Will the lighting be sufficient to support reading? If on a small screen, will the question (and potentially the response options as well) fit without having to scroll? Are there any literacy issues with the population that suggest listening to questions might be easier to comprehend than reading?

Capturing

When deciding how to capture participants' responses to questionnaires, you have to make choices about the types of responses to record, how to record responses, and what you need to know about the timing of responses.

Types of responses to record

One of the first decisions to make is what types of responses will be required. Experience sampling questionnaires can capture a broad range of qualitative and/or quantitative responses—what's important is that you balance the needs of the study with the burden on the participant, as well as with what the experience sampling tool you're using can support. It's not uncommon for experience sampling studies to use any, or often several, of the following question types:

- multiple choice, single answer (e.g., radio buttons);

- multiple choice, multiple answers (e.g., checkboxes);

- scale or rating (e.g., Likert scales, star ratings);

- open-ended, short response;

- open-ended, long response;

- open-ended, numeric response; and

- media capture (e.g., speak the response; take a photo or shoot video of something about the current context).

How to record responses

Once the types of responses that are needed have been determined, you can choose the best way for the participant to record their responses. For example, in the case of questions that call for open-ended responses, would it be more appropriate for the participant to write or type their reply, or to speak it out loud? The answer may depend on the context in which the participant is responding to the questions. For example, are they likely to be in a situation where speaking out loud is inappropriate or will draw unwanted attention to them? Or will speaking be easier than typing or writing a reply? Spoken responses often result in longer and more rich responses, but written replies can be easier to analyze (e.g., there's less opportunity to introduce transcription error or be impacted by environmental noise). It's also possible that instead of (or perhaps in addition to) speaking or writing a reply, having the participant capture a still photo or video might be a good way for them to respond to a question.[28]

---

[28]   As above, check with your institution to see if and how you need to obtain consent from other parties whose data might be captured, or if you need to instruct participants to not include identifiable sound or images of others in their audio, photos, or video.

Timing of responses

You also need to decide what you need to know about the participant's response time, or if you need to set a timeout for questionnaires. For example, it may be helpful to know how long a participant took to respond to the original alert and/or to any of the questions in the questionnaire. It is also usually a good idea to set a time-out for the questionnaire. For example, if the participant hasn't responded to an alert within a certain (usually short) amount of time, you might want the alert to disappear and log that questionnaire as being missed. Similarly, if the participant hasn't completed a particular question or the questionnaire itself within a certain amount of time, you might want to cancel that questionnaire, and log it as started but not completed. In the case of snoozing alerts, it may be necessary to set a maximum number of "snoozes" or a cut-off time at which point that particular questionnaire will no longer be available and will be counted as missed.

4.3.3   LIMITATIONS

While experience sampling collects data that are very helpful for understanding the thoughts, feelings, actions, and experiences that people have out in the world, and it minimizes common limitations of other self-report methods, it is not without its own limitations. In fact, some of its biggest strengths are also its biggest weaknesses. For example, recall and observer bias are minimized with experience sampling because participants respond to questionnaires throughout their everyday lives in the actual context of interest to the researchers. This means that experience sampling interrupts people in their natural environments as they live their lives. For some participants, this can be inappropriate. In our experience, experience sampling can be particularly inappropriate—especially if time-based alerting is being used—for teachers, people who work with clients (e.g., attorneys, doctors, or aestheticians), or people who are not supposed to use their phones or other electronic devices while working (e.g., sales clerks, construction workers, or people who work on film sets) or at school (e.g., children). It can also be disruptive for the people with whom the participants interact. Event-based alerts may be more appropriate for these populations, depending on what the event of interest is and when and how often it occurs (e.g., if questionnaires will trigger only when participants are performing a particular action and they're unlikely to do that while with a client or in an otherwise inappropriate situation, experience sampling may be fine for them). To mitigate this issue, during the recruiting phase, we try to be very clear about the type of commitment we're expecting from participants (i.e., before a potential participant commits to participating in the study, we try to give them a very realistic sense of what to expect).

Experience sampling, as with any measurement, can also potentially impact the very thing you are trying to study, feasibly more so than other forms of measurement because of its very context-dependent nature. Because experience sampling often asks people about their thoughts, feelings, actions, or experiences, they are necessarily reflecting on their thoughts, feelings, actions, or experiences. Such frequent self-reflection can and often does influence their attitudes and behaviors (a phenomenon called *reactivity of self-monitoring*, which is an extension of the *Hawthorne effect*). In our experience, it is critical for the research team to pilot their experience sampling questionnaires in the field, look for any signs of reactivity of self-monitoring, adjust questionnaires as necessary, then pilot again. This may mean holding off on asking participants certain questions until the end of the study (e.g., in an exit interview). We usually plan for pilots lasting at least five days, including over a weekend, with pilot participants who are reasonably representative of the target participant population and are not part of the research team. Below, we discuss an example of when we uncovered reactivity of self-monitoring during a pilot study and had to substantially change our experience sampling and exit interview questions as a result.

Finally, if the research team is loaning equipment to participants (e.g., smartphones or cameras), the equipment may not be returned in its entirety, it may not be returned in working or reusable condition, or it may not be returned at all. We have experienced all of these issues. For example, in cases where we've loaned participants smartphones, it is often the case that the chargers or cases aren't returned to us, that the phones are returned noticeably scratched or even with shattered screens, and in some—fortunately rare—cases, participants have failed to return the equipment at all. Of course, these potential issues are true any time you loan equipment to participants—not just for experience sampling studies.

4.3.4   CASE STUDIES

We frequently use the experience sampling method in our field studies. We have used it for a variety of reasons from trying to better understand how participants use Wi-Fi networks, to how they would want to respond to requests for their location, to helping them troubleshoot problems with research prototypes of wearable computing, and more. In this section, we provide brief overviews of how we used experience sampling in three specific projects. The first is a study that we conducted at Intel Labs Seattle that explored how people use Wi-Fi networking; it illustrates how we had to substantially change our approach to the experience sampling phase of that study when we uncovered reactivity of self-monitoring during our pilot testing. The second is an early-stage study that we also conducted at Intel Labs Seattle that investigated how people would want to respond to requests for their location from members of their social network. Finally, the third is a field study that we conducted at Intel Labs Seattle and the University of Washington of the UbiFit system, where we used experience sampling in a less traditional way to develop UbiFit's journal and help participants troubleshoot problems with the wearable sensing device that detected physical activity.

Understanding of Wi-Fi

In Summer 2008, we conducted a four-week long exploratory field study with 11 participants to investigate people's understanding of laptop computing Wi-Fi use (Klasnja et al., 2009). We were particularly interested in learning what people from the general public who regularly used Wi-Fi understood about it, what their Wi-Fi-related privacy and security concerns were, and what practices they employed to protect themselves from perceived risks. Our three-phased study started and ended with an in-person interview with four weeks of experience sampling and logging on the participants' own laptops in between. Participants were recruited from the general public by a market research firm.

To administer the experience sampling questionnaires, we built a browser plugin that participants downloaded onto their laptops. We used system-triggered, event-based alerts to trigger questionnaires. For example, when the participant connected to a Wi-Fi network that was not in their preferred network list, we triggered a questionnaire and asked several contextual questions (e.g., where they were, what they were doing, and how important the task they were doing was). They received no more than 10 questionnaires per day, but most participants received fewer.

Thanks to pilot testing, we substantially modified our questions for the experience sampling and exit interview portions of the study. Originally, we planned to ask more privacy-related questions during the experience sampling. Although we were careful not to ask about privacy specifically (at least, we thought we had been careful), during piloting, we realized that our questions made pilot testers aware of issues they had never before considered or not thought about carefully, and this new awareness often led to a change in their behavior. This was problematic, as we wanted to *learn* about current behavior, not alter it. For example, one of our original questions asked pilot testers who provided the Wi-Fi network they were using. Such questions made them stop and reflect about that issue—something they had not previously done. This reflection often turned into concern (e.g., Who *is* providing the network? How *do* I know they are who they say they are?), and this concern often turned to behavior change (e.g., I better not use this network after all—I'll just wait until I get home).

In this case, had we not piloted our questionnaires and realized that they were causing reactivity of self-monitoring, we likely would have had to completely repeat the study. We always pilot our instruments (and then revise them and repilot as necessary), especially our experience sampling questionnaires, and in this case, that piloting turned out to uncover a critical flaw with our original plan.

Location Disclosure to Social Relations

In July 2004, we conducted an exploratory study to understand whether and what people were willing to disclose about their location to members of their social network (Consolvo et al., 2005). The 16-participant study started with a survey and an in-lab session that included exercises to identify the people in the participants' social networks, as well as the rules that they thought they would want to apply to requests for their location from those people. The in-lab session was followed by two weeks of experience sampling where they responded to questions about hypothetical requests for their location from individuals in their social network and a nightly voicemail diary study to report if they went anywhere atypical that day. The study concluded with another in-lab session that included an interview, survey, and the opportunity to edit their "rules" from the first in-lab session.

For the experience-sampling portion of the study, we loaned participants Palm m500 devices that were running our iESP software. Participants received 10 randomly timed questionnaires every day for two weeks from 9am–9pm on weekdays and 10am–10pm on weekends. Each questionnaire took approximately 2–3 min to complete and asked questions such as (see Figure 4.2):

- Are you...
  - inside?
  - outside?
  - in a vehicle (end questionnaire)?
- Where are you?
- What are you doing?
- Assume [name X from social network] wants to know your location right now. Would you want to tell [him or her] *something* about your location or *nothing*?
  - Something
  - Nothing
  - (if Something) If the system automatically sends <name X from social network> a response, which of the following level of detail would you want the system to tell [name X from social network] about your location?
    - Exact address
    - Cross streets
    - Neighborhood name
    - Generic place name
    - Zip code
    - City
    - State
    - Country

   ◦ (if Nothing) If the system automatically sends <name X from social network> a response, which of the following would you want the system to tell <name X from social network> about your location?

   • "Request denied"

   • A lie

   • "System busy"

   • "I am busy"

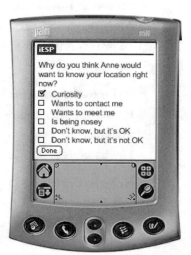

Figure 4.2: Excerpt from an experience-sampling questionnaire. Questions from the experience-sampling portion of the location disclosure to social relations study, as seen on a Palm m500 device running our iESP software.

Customized questions and response options

One of the major changes we made to our iESP software after our prior study on people's information needs (Consolvo and Walker, 2003) was that we added support for something that we called *substitutions* in questions and response options. For example, all of the participants in a study might get the following question:

   Question 100: *Are you with <X> right now?* (yes)   (no)

Where for Participant 38 (or "P38"), X=Mary; for P39, X=Peter; and for P40, X=Paul. The same applied to response options. For example:

   Question 200: *Which of the following do you currently have with you?*

○ <A1>

○ <A2>

○ Cell phone

Where for P38, A1=black leather purse and A2=rose gold watch; for P39, A1=silver backpack and A2=GPS watch; and for P40, A1=fuchsia waist pack and A2=calculator watch. All three participants would see the option "Cell phone." These substitutions also meant that the research team could:

- **Pick one substitution from a list at random.** Assume that P38's friends include Mary, Jane, and Jill. Each time P38 was asked about who they are with, iESP could be programmed to pick one of those three names at random to substitute in a given question for P38.

- **Refer to the previous substitution in a follow-up question.** Assume that iESP chose to ask P38 about Jane in Question 100 and P38 responded "no." iESP could be programmed to follow-up with a question that asked about Jane, for example, "*Do you plan to see Jane later today?*" iESP would correctly choose Jane, instead of Mary or Jill.

- **Refer to the previous question's selected response in another question.** Assume that P38 responded to Question 200 that they currently have their rose gold watch with them. iESP could be programmed to follow-up with a question specifically about the rose gold watch.

Prior to this study, related work on people's location disclosure preferences tended to be done via standard one-time surveys and ask about how participants would want to respond in general to categories of people from their social network (e.g., *what would you want to tell a* <friend, coworker, family member> *about your location?* rather than what *would you want to tell Jane about your location?*). Data from prior work was also collected out of the context in which requests would likely be made.

In this study, because we used experience sampling rather than traditional one-time surveys, we were able to collect data about how participants would want to respond to requests for location in a variety of the types of contexts those requests would actually occur. We found that context often mattered. For example, even though they would usually be happy to tell their spouse where they were, if they hadn't left the office yet but they were supposed to have left, they might want to tell their spouse that they were already on their way home. Or though they wouldn't usually want to tell their boss where they were, they might be happy to do so in the midst of an important work deadline. And because of the substitution functionality we added to iESP, we were able to ask about multiple individuals from various categories of people from participants' social networks. Unsurprisingly, we found that participants often wouldn't want to treat everyone from the category the same way (e.g., they might want to respond differently to different co-workers or to different friends or family members). While these results were arguably unsurprising, they were previously unconfirmed.

Incentives and compliance rates

The incentive for participation in this study was up to $250 USD and was determined based on level of participation. The total number of experience sampling questionnaires completed factored heavily into how we calculated the incentive, which was paid at the end of the final in-lab session. Incentives were:

- $___ for participating in the two in-lab sessions, returning the equipment, and completing ___% of the 140 experience sampling questionnaires as follows:

  - $60 for completing 50%;

  - $85 for completing 65%;

  - $175 for completing 80%;

  - $225 for completing 95%; and

  - $250 for completing 99-100%.

Our average response rate for the experience sampling portion of the study was 90.4%, with 12 of 16 participants having a response rate of 95% or higher. The only participant with a response rate of less than 74% had a rate of 42%; they explained that a family emergency came up during the study that substantially impacted their ability to participate. Compared to our prior investigation of people's information needs and what output devices were available to them throughout the day (Consolvo and Walker, 2003), this was a big improvement in response rate. In the information needs and output devices study, participants similarly received 10 randomly scheduled alerts each day, but only over a one-week period. Each time they were alerted, participants were asked to use a separate digital camera to take a photo of what they were doing, in addition to responding to the questionnaire on the PDA.

The incentive for participating was up to $120 USD, paid at the end of the final in-lab session, as follows:

- $50 for participating in the 2 in-lab sessions *and* returning the experience sampling equipment, and

- $1 for each of the 70 experience sampling questionnaires that were completed (i.e., max of $70).

We had an 80% average compliance rate (i.e., on average, each participant completed 56 of 70 possible questionnaires; median: 58; range: 20–68). The average number of photos taken (which should have been a 1:1 match with the questionnaires), was 52 (median: 56; range: 2–68). In the exit interviews, participants told us that their reasons for not completing questionnaires was either that it would have been inappropriate to respond in their current situation or that they simply didn't notice the alert. While we cannot say for certain that the different incentive structures caused the differences in response rate between the studies, we suspect that it was a strong factor. That said, as above, you should check with your institution on what type of incentive structure is most appropriate for your study.

UbiFit

UbiFit is a mobile technology that we developed to encourage people to incorporate regular and varied physical activity into their everyday lives (Consolvo et al., 2008a, 2008b, 2014). The system included a mobile phone (i.e., a Cingular 2125 from AT&T Wireless with Windows Mobile 5.0) that ran a glanceable display on the phone's wallpaper that represented key information about the user's physical activity and goal attainment, and it also included a journal where the user could manually add, edit, and delete information about their activities and monitor their progress toward their goal. The phone was accompanied by a separate fitness device that automatically inferred and communicated information about certain types of physical activities to the UbiFit application on the phone. The UbiFit system that we built was a research prototype (i.e., it was not available commercially), and we evaluated it in two field studies—a three-week study in Summer 2007 and a three-month study over the 2007–2008 Winter holiday season—with participants who were recruited from the general public by a market research firm.

We used experience sampling in two somewhat non-traditional ways in the UbiFit field studies. First, the UbiFit journal on the mobile phone was built using Froehlich et al.'s MyExperience tool (Froehlich et al., 2007), which was mentioned earlier. To log an activity in the journal, the participant essentially completed an experience-sampling questionnaire. The participants could either initiate entering the activity themselves (i.e., a participant-triggered, event-based alert), or they could respond to a reminder that we set up to ask if they had anything to add to their journal. The system-triggered reminder was only triggered if the participant hadn't added anything to their journal in a certain time frame (i.e., a system-triggered, event-based alert), so if they were actively journaling, they wouldn't be bothered with a reminder. Not only did MyExperience give us a reasonably quick way to build the journal and intelligently manage reminders, but it also provided a way to elegantly handle interruptions on the phone (e.g., if the participant received a phone call while in the middle of journaling an activity, MyExperience allowed the participant to respond to the call, then return to where they left off in the journal /"questionnaire").

The second way we used experience sampling was to troubleshoot problems with the fitness device during the field studies. UbiFit's fitness device was the *mobile sensing platform* (Choudhury et al., 2008), a wearable sensing platform developed by Intel Labs Seattle and the University of Washington that we trained to automatically detect walking, running, cycling, using the stair machine, and using the elliptical trainer. The mobile sensing platform sent activity predictions to the UbiFit application on the phone several times per second. As with the UbiFit system, the mobile sensing platform was a research prototype. If the UbiFit application on the phone lost its connection with the mobile sensing platform, we used MyExperience's system-triggered, event-based alerts to trigger a questionnaire for the participant to answer. That questionnaire tried to diagnose the reason for the lost connectivity and alert the principal investigator if the problem turned out to be a device failure. For example, the questionnaire started by asking if the participant had the fitness device with them. If they replied that it wasn't with them, the questionnaire asked them why they left it behind, then thanked them for their time and logged the reason for the lost connectivity. However, if the participant said that the fitness device was with them, we asked other questions in an attempt to diagnose the problem. For example, were the phone and device near each other? Was the device's power light on? Was the device charged? If the power light was on and the device and phone were near each other, then the principal investigator received a text message alerting her to the problem so that she could replace the device. In this case, experience sampling enabled us to troubleshoot technology problems in the field, collect data about the situations in which connectivity between the system's two pieces of mobile technology was lost (including helpful details such as a participant purposely leaving the device behind because it didn't look good with their outfit), and replace faulty equipment in a timely fashion, all with minimal effort from participants.

4.4   SUMMARY

In this chapter, we discussed two self-report methods—diary studies and experience sampling—that can be helpful for understanding the thoughts, feelings, actions, and experiences that people have out in the world. With *diary studies*, participants keep a log (often on a daily basis), and with *experience sampling*, participants complete brief questionnaires (often several times per day) as they go about their everyday lives. Both are self-report methods that are conducted in the field, typically for one week or longer, and are often combined with other data collection methods, such as interviews or usage logs, as part of a larger study. These methods have many similarities and can be used to collect similar types of data (including qualitative and/or quantitative). Perhaps the chief difference is in their timing. Although both collect data often, participants tend to provide data at a time that's convenient to them in diary studies, and in the moment, often in response to an alert, in experience sampling studies. As you can see from the descriptions above, the line separating the methods is blurry. Both can provide very helpful data and insights. Which to use depends on the needs of the research and the population being studied, and it might be appropriate to use a hybrid of both (similar to what we did in Matthews et al., 2016).

CHAPTER 5

Answering "Did it work?": A Primer to Experimental Designs to Test for Change

"Did it work?" This chapter discusses how to answer that seemingly simple question when your mobile system is intended to promote change (as opposed to simple observation, which is discussed in Chapters 2 and 4 on sensing and experience sampling). As you will soon see, answering that question is far more complicated than it might seem.

Take, for example, a mobile health (mHealth) intervention to change behavior such as UbiFit (Consolvo et al., 2008b, 2014; discussed in other chapters). While the focus of the UbiFit app was to promote increased physical activity, it is difficult to determine if it was the factor that actually promoted change, and even if it did change behavior, it is unlikely that it changed ONLY physical activity. As a person tries to be more physically active, several other factors might also start to change such as their daily routine (e.g. waking up earlier to go to the gym before work), their mood, their eating might improve, and, perhaps, they may be exercising with friends and thus have increased social contact with others. Not all of these changes will occur with every person however (e.g., not everyone will try to fit in physical activity in the morning), nor will they occur at the same rate of change (e.g., some might immediately become more active when first using an intervention, whereas others might have more of a ramp up). Further, while your mobile system may start the process, it may end up being some other factor that perpetuates the change. While you might be thinking, "who cares *how* it works, I just want to know *if* it works;" answering the "how" question leads to better designs as it results in knowledge that you can use in future design challenges far more readily than just knowing "if" it worked.

The purpose of this chapter is to provide you with a primer for understanding how to pick the right experimental design for helping to evaluate both, if your mobile system "works" and, ideally, "how" your system might be promoting change. We start with some basics about experimental design, including describing the building blocks for parsing out a cause and effect relationship, defining an experiment that is used to determine cause and effect relationships, and establishing some of the competing assumptions involved when designing an experiment. Following this, we will provide a high-level summary of the myriad experimental designs that a mobile user researcher could feasibly use within their studies. Note that our purpose is to provide you with just enough information to help you choose the right experimental design for your particular problem, not to provide enough information to confidently use the experimental designs. For this latter purpose, we will direct you to seminal papers and tutorials to learn the methods and studies that have used each experimental design.

5.1   ESTABLISHING CAUSE AND EFFECT: SCIENCE 101

As most mobile user research endeavors are focused on humans and the interaction between humans and digital systems, we will be building on the scientific traditions from the social and behavioral sciences. While we only discuss this at a high-level, interested readers in the behavioral scientific method should consult classic works on this topic such as Popper (2005), Fisher (1925), Kuhn (2012), Cook and Campbell (1979), Shadish et al. (2002), and Cronbach (1957). For the sake of this chapter, we rely on the tradition of establishing confidence in cause and effect relationships (i.e., causal inference) set up by Campbell and Stanley (1966), which was most recently updated by Shadish et al. (2002). If you are well versed in the behavioral scientific method, you can skip this section and go straight to the next section: "Primer of Experimental Designs."

A high-level purpose of science is to describe, predict, and control the physical and social world. Prediction and control is achieved by establishing generalized causal inferences, achieved through three parallel strategies:

1. **Observation:** establishing that observation takes precedence over theory for defining truth (e.g., think Galileo making observations that the Earth was not at the center of the universe; observation took precedence over the "known fact" of the geocentric theory of the time);

2. **Manipulation:** manipulating some facet of a system (e.g., an intervention to change behavior), followed by systematically observing the target outcome to see if it changes; and

3. **Control:** utilizing a variety of strategies such as improved measurement or random assignment to control for extraneous "third variable" factors that might be explaining an observed relationship between two variables of interest.

At a general level, causal inference can be achieved in a variety of ways as long as the three strategies of observation, manipulation, and controlling for other plausible causal factors are used in tandem. At the most fundamental level then, *an experiment is any study that utilizes observation, manipulation, and control in tandem for inferring cause and effect relationships.*

Note, this is in contrast to other types of studies that do not include all of these elements such as cross-sectional observational studies. Such studies observe phenomena but not over time, which is a pivotal "third variable" factor necessary to establish a logical argument of cause-and-effect; conceptually, a cause should temporally occur before an effect. They also do not involve any manipulation. For example, a mobile user researcher might be interested in determining if health and fitness apps "work" to promote increased physical activity. To test this, the researcher may conduct a survey asking individuals about the fitness tracking apps they currently use and also their current physical activity levels, and then explores if there is a correlation between the two self-report measures. While valuable, this type of study provides little insight on what factor is *causing* the other as a fitness tracking app may impact physical activity but, perhaps those individuals who are physically active tend to use fitness apps. It could also be plausible that some third factor (e.g., being a college student or not) may actually be the key factor driving the relationship between the two other variables. While doing observational studies over time (i.e., *longitudinal studies*) can help to establish the time-ordering precedence (e.g., measuring physical activity, then measuring app use at a later time, and then seeing an increase in physical activity after that), a lack of manipulation provides only limited control for alternative "third variables" or extraneous unmeasured factors that could explain the causal inference. While we do not discount the value of these other methods (and, indeed, Chapter 4's discussion on experience sampling is a logical extension of the longitudinal design work that is very well suited for mobile user research), they are not the focus of this chapter. In this chapter, we focus on *experiments designed to infer causal inference.*

There are often two categories used to distinguish experiments: (1) those that include random assignment or not (classically labeled *experimental vs. quasi-experimental,* respectively), and (2) whether the study is used primarily to infer a causal inference on a more individual-level (i.e., classically labeled "*within-person,*" meaning that a statement can be made about how variations that occur within individuals on one variable can be used to predict variations on another variable over time[29]) vs. a study used to infer causal inference, on average, across individuals/groups of individuals (i.e., classically labeled, "*between-person*" meaning that a statement can be drawn about an effect about a collection of individuals/at the group level). These categories are important as they require different logical assumptions and, ultimately, are asking different research questions.

Table 5.1 highlights just some of the experimental designs that are used within the social and behavioral sciences to infer causal inference. As this table illustrates, there are a wide range of experimental and quasi-experimental designs available to establish a cause-and-effect relationship. This table is actually limited in scope as many of these design strategies can be combined and used in tandem. For example, it is plausible to combine a between-*person* randomized controlled trial strategy of randomly assigning individuals to different groups and to then add another level of randomization using one of the within-*person* designs. Further, experimental designs can be used in novel ways to answer several different types of questions. For example, a factorial study design could answer the research question, "what is the *best* intervention combination package for promoting behavior change?" An alternative research question though could be, "which intervention components are most useful for promoting change?" While they might seem similar and can be answered with

---

[29]   Technically, most within-person designs are created to infer "on average" effects across individuals over time, not develop a causal inference on a case-by-case basis. For more information on this, see Hekler et al. 2016a.

the same study design, they require radically different resources (the second one is *much* easier to conduct and is often "good enough"). For example, a 2x2x2 factorial design could be designed to test out the components of an intervention to help someone eat better and include the following factors:

- **factor 1:** providing goal-setting or not;

- **factor 2:** positive reinforcement or not; and

- **factor 3:** social support or not.

To answer the first question, the study would need to include enough participants to test the three-way interaction between the components, but the latter question only requires just enough users to test for the weakest of the three interventions being tested (more on this below when we discuss the "MOST" strategy). As such, the latter question can often be "good enough", and the study you would need to run to answer it would be much easier to conduct compared to investigating the former question about the "best" intervention. At the conclusion of this chapter, our goal is to provide you with a high-level understanding of these sorts of issues for experimental designs that are valuable specifically for mobile user research.

Table 5.1: High-level types of experimental and quasi-experimental designs

Experimental Design		Quasi-Experimental Design	
Within-Person	Between-Person	Within-Person	Between-Person
Alternating Treatment Design	Randomized Controlled Trial	Pre-Post Comparison Design	Non-assigned group comparison
Micro-randomization design	Cluster-randomized controlled trial	Removed or Repeated Treatment Design (also known as Reversal or ABA)	Cluster non-assigned group comparison
System ID Informative Experiment	Factorial/Fractional Factorial Design as used within the Multiphase Optimization Strategy (MOST)	Interrupted Time-Series Designs	
	Sequential Multiple Assignment Randomized Trial (SMART)		
	Continuous Evaluation of an Evolving Behavioral Intervention Technology (CEEBIT)		

We also highlight all of the designs in Table 5.1 to acknowledge the impressive number of strategies for achieving effective observation, manipulation, and control. Unfortunately, most mobile user researchers are only aware of a very restricted range of designs such as the pre-post comparison, randomized controlled trial, or quasi-experimental design of comparing naturalistic groups.

The goal of this chapter is to recap these designs but then also introduce you to other experimental designs that are particularly useful for mobile user researchers. There are three reasons why this can be valuable. First, it is, unfortunately, far too common among mobile user researchers (and indeed the behavioral science community as well) to fall back to "tried and true" research methods because they are aware of them, not because the method is appropriate. For example, you may decide to use a between-person randomized controlled trial to test if your system is better than a no-treatment control but what you are really interested in is understanding how varying interactions over time with your system influenced a target outcome (which you can't infer from a randomized controlled trial). Second, providing you with a primer of these other experimental designs can provide you with a framing to expand the range of questions you may be inspired to ask. For example, a mobile user researcher may believe that a factorial design is far too difficult to conduct, even if the core question they are interested in is understanding how each component of a system works to promote change. Knowledge about ways to reuse this experimental design (i.e., focusing only on main effects) can be a highly efficient way of testing a variety of interesting research questions in tandem and, indeed, would allow for the study of the effects of the components to be examined. Third, with this primer, our goal is to provide you with just enough information to identify plausible experimental designs for your mobile user research problem and to know where to look next for more information. This last point is critical to help you move beyond simple awareness to actually using the methods in your work.

In line with being a "primer," we organize our discussion around the most valuable research question(s) you can adequately answer with each design (though note that methods often can be feasibly used to ask several different research questions as our factorial design questions illustrate; we hit on the type of question(s) a mobile user research might want to ask with each experimental design). While the distinctions are sometimes quite subtle, a high-level understanding of these subtleties will enable you to properly ask the question you have and then match it to the right method.

5.2   PRIMER OF EXPERIMENTAL DESIGNS

With these points established, we now turn to a whirlwind primer of experimental designs. For this, our goal is to give you insights on when you might want to use one over another. As such, our review remains highly pragmatic but will ultimately need to be supplemented with additional resources on designing the study and conducting the appropriately matched statistical analyses and possibly training depending on the design.

Experimental Designs	Core Research Question	What's being "controlled"?	When to Use
Table 5.2: High-level types of experimental designs			
Basic Within-Person Quasi-Experimental Designs			
Pre-Post Comparison	Was there a change over time in my target outcome comparing times prior to implementing my system to after implementing?	Establishing temporal precedence and using manipulation but otherwise limited control beyond "statistical" control	Use when no other designs are possible.
Interrupted Time-Series (i.e., Removed/ Repeated Treatment Design, also sometimes called ABA)	Does the system result in a significant change in a core target outcome that dissipates after removal?	Controls for natural changes that could have occurred through time alone.	Use when it is possible to have the effect of the system "dissipate" when it is removed.
Interrupted Time-Series Designs (i.e., Multiple Baseline)	Was there a change over time in my target outcome comparing times prior to implementing my system to after implementing?	Phased baselines across participants allows for control of time-varying "third variable" factors	Use this if you are interested in testing for change and if randomization of individuals or entities is difficult.
Between-Person Quasi-Experimental Designs			
"quasi-experimental" design	Does the system produce change in the target outcome relative to a non-randomized control group?	Controls for natural changes that could have occurred through time alone.	Use when randomization is not possible AND a meaningful comparison group exists.
Between-Person Randomized Experiments			
Randomized Controlled Trial	Does this entire system package result in a significant change in a target outcome relative to control?	Controls for "third variable" factors that were not measured between groups.	- When the research question is about the entire system (e.g., quality assurance). - When randomization is possible - When enough resources are available to conduct the study

Experimental Designs	Advantages/Disadvantages
Basic Within-Person Quasi-Experimental Designs	
Pre-Post Comparison	ADV: - Easy to implement DIS: - Causal inference suspect as only limited control used.
Interrupted Time-Series (i.e., Removed/Repeated Treatment Design, also sometimes called ABA)	ADV: -Easy to implement -Stronger causal inference compared pre-post only DIS: -Can be hard to "remove" a system
Interrupted Time-Series Designs (i.e., Multiple Baseline)	ADV: - An elegant strategy for controlling for time-effects between participants that does not require a control group or the removal of the system DIS: - Technically not a within-subject comparison as the control is achieved via different phased baselines across participants - Similar to an RCT or other between-participant designs, will likely require more participants to achieve power compared to more classic within-participant designs - Particularly vulnerable to possible novelty effects of an intervention
Between-Person Quasi-Experimental Designs	
"quasi-experimental" design	ADV: -Allows for improved control of time when randomization or removed/repeated design used DIS: - Selection of the control group greatly impacts the confidence in the conclusion
Between-Person Randomized Experiments	
Randomized Controlled Trial	ADV: - Elegant mechanism for controlling for "third variable" factors. DIS: - Often highly resource-intensive to do properly - Limited insights can be drawn for specific individuals. - Limited insights drawn about the "active ingredients" components of intervention including design decisions

Table 5.2 (continued): High-level types of experimental designs

Experimental Designs	Core Research Question	What's being "controlled"?	When to Use
Between-Person Randomized Experiments (continued)			
Factorial/Fractional Factorial Design as used within the Multiphase Optimization Strategy (MOST)	Primary question: What is the independent impact of each component of a system on impacting a target outcome? Secondary question: What is the interactive effect of intervention components?	- Controls for "third variable" factors that were not measured between groups. - Controls for interactions between intervention components (and, indeed, can examine this uniquely via secondary question	- Use this when you are really interested in the impact of the components of your intervention. -Use this if you want to develop an optimized intervention.
Sequential Multiple Assignment Randomized Trial (SMART)	- What is the appropriate decision rule to use when a person does not respond to my first intervention?	- Controls for "third variable" factors that were not measured between groups. - Uniquely controls for the main effect of an intervention and takes into account decisions that would be made for those not responding.	Use when you want to know if your decision is right for non-responders to an intervention (e.g., if they aren't losing weight, I'll try this new intervention)
Within-Person Experimental Designs			
Micro-randomization design	Primary Question: - What are the proximal and delayed effects of an intervention component? Secondary Questions: - How do the proximal and delayed effects of an intervention component change over time? - Which factors (fixed or time-varying) moderate an intervention component's proximal or delayed effects?	Allows for within-person insights to be drawn by controlling for changing delivery of interventions over time	- When it is feasible to actively activate and deactive components of a system over time AND the expected duration of the effect can be observed relatively quickly (i.e., prior to the next delivery of the alternating component) - Unique compared to classic alternating treatment designs, allows for multiple components to be tested simultaneously
System ID Informative Experiment (Sub-type of micro-randomization study)	What is the mathematical model that could be used to guide decisions on when, where, and how best to deliver intervention components?	Allows for within-person insights to be drawn by controlling for changing delivery of interventions over time	When the focus is on building mathematical models (i.e., algorithms) for guiding future decision making by a system.

Experimental Designs	Advantages/Disadvantages
Between-Person Randomized Experiments (continued)	
Factorial/ Fractional Factorial Design as used within the Multiphase Optimization Strategy (MOST)	ADV: - Elegant mechanism for controlling for "third variable" factors. - Highly efficient design that allows multiple questions (i.e., evaluation of each component) to be answered simultaneously. - Provides a structure for examining interaction between components. -Elegant strategy for controlling for "third variables." DIS: - Can be confusing to use without prior training, particularly if "fractional factorial design" used. - Limited insights can be drawn for specific individuals.
Sequential Multiple Assignment Randomized Trial (SMART)	ADV: - Elegant mechanism for controlling for "third variable" factors. - Uniquely useful for testing the utility of decisions made within an intervention based on the response individuals have to the intervention. DIS: - A complicated design that requires appropriate training to use, particularly the statistics. - Focus is still on the impact "on average" between individuals; does not give insights on a case-by-case basis.
Within-Person Experimental Designs	
Micro-randomization design	ADV: - Elegant within-person design that provides strong internal causal inference at an individual-level. - Highly resource-efficient from a user perspective as inferences can be drawn with far fewer insights. DIS: - A fairly complicated design that requires good knowledge of statistics to properly analyze the data.
System ID Informative Experiment (Sub-type of micro- randomization study)	ADV: - methods provide a highly refined strategy for parsing out signal effects from time (e.g., cyclical patterns, "main effect" responses, and transient responses). DIS: - Details still being worked out and requires advanced understanding of system identification methods of analysis (e.g., dynamical systems modeling) including knowledge about uses of differential equations.

Experimental Designs	Core Research Question	What's being "controlled"?	When to Use
Table 5.2 (continued): High-level types of experimental designs			
Emerging Experimental Designs			
Continuous Evaluation of an Evolving Behavioral Intervention Technology (CEEBIT)	Are the newer versions of a system resulting in significant improvements relative to older versions of the system?	Uses past versions of a system as a defacto control, thus allowing for a clear test of the value of new features added to a system	When the primary goal is to have an evidence-based strategy for facilitating continuous quality improvement of an app

5.2.1 WITHIN-PERSON QUASI-EXPERIMENTAL DESIGNS

Pre-Post Designs

The *pre-post design* is, in many ways, the starting point for most of the other designs. This design involves measuring an outcome of interest, introducing some kind of manipulation (e.g., provide individuals with a mobile app designed to elicit change), and then observing the outcome again to see if there was any change compared to the pre-phase. This design allows you to answer the question, "Was there a change over time in my target outcome from pre-implementation to post-implementation of my system?" Measuring the outcome pre- and post-implementation is an essential first ingredient for an experiment as it provides the temporal precedence (arguably the most important first factor for achieving control) and also uses an active manipulation. As such, *it establishes the bare minimum of observation, manipulation, and control.*

The key advantage to this study design is that it is very easy to use and implement. The problem, however, with this design is that the causal inference is highly suspect as there is only the bare minimum level of control. For example, it is plausible that something else occurred at the same time a person received an app, and other factors (such as interactions with the research team) ultimately produced a measured change. Further, this study design is highly susceptible to the "novelty effect," which is the tendency for a change to occur in the desired outcome whenever a new technology is introduced, not because of any actual value in the system to produce the effect. In other words, there was an observed effect because a "new toy" was present to play with. As this is such an incredibly common phenomenon with any mobile user research project, it renders this design incredibly suspect unless the study can last for a "long enough" duration to allow any novelty effect to dissipate (this is best determined on a case-by-case basis, depending on the problem, but in our work, we have found that a novelty effect can sometimes last as long as 8 weeks).

We establish this starting point method to state that it is a plausible experimental design but likely should only be used if no other option is available (e.g., highly limited resources, including no experience with the other experimental designs). From a statistical analysis standpoint, this experimental design requires only beginner-level knowledge of statistics as a rudimentary sense

Experimental Designs	Advantages/Disadvantages
Emerging Experimental Designs	
Continuous Evaluation of an Evolving Behavioral Intervention Technology (CEEBIT)	ADV: - Highly elegant design that fits well with current development practices. - Provides a logical mechanism of control. DIS: - Still not fully developed or tested to ensure insights are correct.

of an effect can often be inferred with simple visualizations (although more advanced statistical techniques would be available and appropriate for increasing confidence). For more information on this study design, please see Shadish et al. (2002).

An example of the use of a pre-post design with no control group is a study by Burns et al. (2011). They tested the feasibility, functional reliability, and patient satisfaction of "Mobilyze!", a mobile and internet-based intervention for the treatment of major depressive disorder through an 8-week single-arm field trial. They developed machine-learning models through the use of context-sensing and ecological momentary assessment to predict a patient's mood, emotions, environmental context, social context, etc. The interventions consisted of ecological momentary interventions, an interactive website for behavioral skills training and email and telephone support from an assigned participant. Clinical outcome measures of major depressive disorder were measured at baseline, 4 weeks, and 8 weeks (the minimum threshold we recommend for allowing a novelty effect to dissipate). As mentioned above, in early stages of development (in this case, piloting the context-sensing system that aimed to identify and respond to patient states), or when resources are limited (recruiting patients with major depressive disorder is highly difficult), a pre-post design can be an appropriate design.

Interrupted Time-Series Design

The core logic of an *interrupted time-series design* is that many observations are made on the same variable over time, and different "interventions" are delivered or not over time. There are two classic ways that an interrupted time-series strategy can be implemented: (1) a purely within-person strategy (e.g., measurement only, followed by an intervention period, followed by a return to measurement only, which is called a *reversal* or *ABA design*; note there are also many variations to this general idea, such as an ABB[1] or as used in n-of-1 trials, (Kravitz et al., 2014)); and (2) a between-person strategy (e.g., using different baseline lengths for different participants and then observing if changes occur for each user at the different baseline period). We will discuss each of these sub-strategies of an interrupted time-series design in turn.

Reversal/ABA Design

The *reversal design* (also known as *Removal* or *ABA design*) uses the same basic logic of a pre-post design but adds the removal of the intervention (e.g., take away the app) followed by continued observation of the target outcome. The research question that is answered with this design is, "Does the system result in a significant change in a core target outcome that dissipates after removal?" Removing an intervention and then examining if the effect goes away is an elegant strategy for increasing confidence in the causal impact of the intervention as the dissipation of the effect increases confidence that the intervention is the key factor influencing an outcome. As such, there is stronger causal inference than the simple pre-post. This study design is also advantageous because it is relatively easy to implement. From a statistical analysis standpoint, this experimental design requires only beginner-level knowledge of statistics to start to infer effects, including simple visualizations (although, again, more advanced statistical techniques would be available and appropriate for refining insights drawn).

The fundamental problem with this experimental design, particularly when the treatment is feasibly removed (although not necessarily the case for other variations, such as those used in n-of-1 trials (Kravitz et al., 2014)), is that it is often rare and undesired to have an effect only be present when a person is actively using the system. Put differently, when change is desired, it is often also desired to have that change "stick" even after someone stops using a system. As such, this design is only appropriate when the desired effect is meant only to be observed when a person uses the system or if the expectation is that the system is truly required to achieve an effect (a point we will emphasize with our example below). Another disadvantage of this design is that, like the pre-post design, it is highly susceptible to being impacted by the novelty effect. For more information on this study design, see either Dallery et al. (2007, 2013), Kravitz et al. (2014), or Shadish et al. (2002).

Raiff and Dallery (2010) utilized the ABA design to assess the feasibility of an Internet-based incentive program to increase adherence to self-monitoring blood glucose levels in four teenagers with Type 1 Diabetes who were performing blood-testing less than four times a day. Each of the three phases (A - baseline, B - treatment, and A - return to baseline) lasted for five days. Data for both baseline (or "A") phases was collected from the participant's personal glucometer. In the treatment phase, participants had to post videos of themselves performing the glucose testing and received monetary incentives for adhering to the recommendations (performing at least 4 tests per day). At the end of the intervention, the participants were asked to discontinue posting videos but to continue blood testing as per the recommendations. The observed mean values for tests per day were 1.7 and 3.1 for baseline and return to baseline phases, respectively, and 5.7 for the treatment phase. The reduction in the number of tests per day after the incentive was removed increases confidence that the internet-based contingency management intervention improved adherence to glucose testing in teens but also that it is needed to continue to facilitate adherence (at least in the short-term).

Multiple Baseline Design

In *multiple baseline designs*, a basic pre-post design is used for each individual or "person" under study (e.g., instituting the use of a new system at an organizational level) but each individual has a different length of baseline implementation. For example, a classic way to use this design is to begin measuring several organizations on an important outcome metric like worksite productivity and then incrementally introduce the manipulation (e.g., a new intervention), with each organization receiving the system at a different time (e.g., after two weeks for organization 1, after four weeks for organization 2, after six weeks for organization 3, after eight weeks for organization 4, etc.). While this study is often labeled a *within-person* design, since it uses multiple different persons as a fundamental strategy for its analysis, it is technically suited to answer the between-group research question of, **"was there change over time in my target outcome from pre to post, controlling for broader changes that occurred during that time-period (e.g., changes in cultural norms or major news stories)?"**

This design is an excellent design to use, particularly when it is not possible to do the reversal strategy described earlier or when it is not feasible to randomize groups and, often, when it can be difficult to find a meaningful comparison group. For example, this type of experimental design is very well suited for mobile user research questions that might involve implementing different systems at different organizations (e.g., progressively rolling out a mobile system to different organizations). From a statistical analysis standpoint, this experimental design requires a beginner-level knowledge of statistics, as an inference can often be gleaned through careful visualization (again though, more advanced techniques are available). For more information on this experimental design, see Dallery et al. (2007, 2013), and Biglan et al. (2000).

Cushing et al. (2010) utilized the multiple baseline design to examine the effectiveness of a personal electronic device (PED) in improving adherence to self-monitoring of dietary intake and physical activity in three overweight adolescents. This was part of a weight management program. The participants were given daily dietary and physical activity self-monitoring goals, and adherence was assessed by calculating percent weekly goal attainment. Self-monitoring was done using the traditional paper and pencil method during the baseline phase. The intervention (utilizing the PED instead of paper and pencil) was introduced in a staggering fashion for all three participants (Figure 5.1). Goal attainment increased for all three participants after the introduction of the PED. This was observed for each participant in spite of the fact that the baseline phases for all participants varied and were introduced systematically in a "staggered" manner. Confidence in the utility of the PED was increased because the change was only observed for each person after the introduction of the PED. Another example is the more recent work, *MyBehavior* (Rabbi et al., 2015).

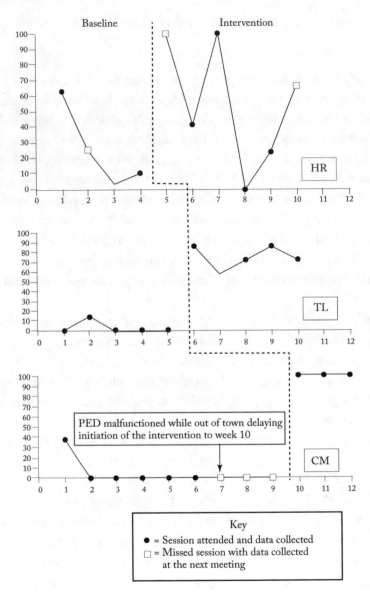

Figure 5.1: Comparison of baseline and intervention percent goal attainment. Dashed lines indicate the transition from the baseline to intervention phase of treatment. Diamonds represent weeks where a treatment session occurred, while square indicate a missed session. From Cushing et al. (2010).

5.2.2   BETWEEN-PERSON QUASI-EXPERIMENTAL DESIGNS

"Quasi-Experimental" Design

Classically, a study design that involves pre-post measurement coupled with a comparison of two naturalistic groups is often labeled a "quasi-experimental" design. It would likely be more accurate to label it a pre-post non-randomized group comparison design, as there are a variety of quasi-experimental designs possible (as per examples illustrated here). The basic strategy is to measure some outcome of interest prior to some manipulation occurring in different groups that can be meaningfully compared (a critical point that we will return to) and then have a manipulation occur followed by continued tracking. For example, we used this quasi-experimental design by comparing the impact of one class that learned about issues related to food and food production and compared their eating habits to three other classes focused on health issues (Hekler et al. 2010). Randomization could not occur as the students self-selected which class they participated in, but pre-post comparisons of metrics compared between these different classes provided some confidence that the class may have incited improved eating behavior among those in the food and food production class. As implied here, this study design is well suited to answer the question, **"Does the system produce change in the target outcome relative to a non-randomized control group?"** This study design is useful when randomization cannot occur (e.g., comparing different self-selected classes or groups) AND meaningful comparison groups can be found. As discussed earlier with regard to the "control group," the importance of a "meaningful" comparison group is a critical concept for all between-unit experimental designs, both quasi-experimental and experimental. A meaningful comparison is ultimately dictated by the mobile user researcher's research question and often is where the real subtleties in terms of good experimental design come into play. We build on this point with the rest of the between-unit quasi-experimental/experimental designs we discuss but, for the purposes of a pre-post non-randomized group comparison design, an essential question to ask is to clearly examine the factors that will be different *beforehand* between two groups.

Returning to the class example, it could be inferred that students who sign up for a class about food and food production might be more interested and motivated by thinking about systems-level issues in society (indeed, the class was called "Food and Society"). In contrast, students that sign up for a health psychology class (one of the comparison classes) may be more interested in personal health and individual-level behavior change. These distinctions matter, as they are differences that are present prior to the manipulation and thus must be taken into account within the study. This can be done through measurement of plausible factors that will differ between groups at baseline and then statistically controlling for those factors. Even then, the manipulation itself (in this study, the content of the two classes) may not alone have driven any observed effect but, instead, the two groups just might have naturally started to differ over time.

This is the fundamental risk of between-person quasi-experimental designs; basically that differences that were present prior to a study will ultimately be the driving force of change. As such, we caution any mobile user researcher in the use of this design. In our view, it is one of the riskier experimental designs for inferring causal inference, mainly because it seems so deceptively simple. From a statistical analysis standpoint, it requires an intermediate level of statistics to establish basic causal inferences, after controlling for differences between groups. Overall, the design can be used effectively but it requires a very careful examination of the questions.

1. What is my research question?

2. What exactly do I need to be controlling to answer my research question? and

3. What am I controlling for when I compare these groups?

For more information about this study design, please see Shadish et al. (2002).

Another example of this type of design is a study by Nundy et al. (2014) that tested the use of CareSmarts, a multicomponent mHealth diabetes program intended to provide self-management support and team-based care management for people with diabetes who were enrolled at the University of Chicago Medical Center. The primary form of intervention was through automated text messages and intervention components were based on constructs such as cuing, self-efficacy, social support, and health beliefs. Participants received text messages about self-care of diabetes, some prompts-to-action and also some questions asking them if they needed refills of any medication. Participants also had to complete a 10-week education curriculum covering various modules related to self-care. In this study, all eligible (above 18 years, and with Type I or II diabetes) employees registered with the center's employee health plan were contacted for enrollment, and those who agreed were analyzed in the treatment group while those who declined were considered as the control group. Post-intervention analyses showed that control of HbA1C (glycated hemoglobin), adherence to medication, self-monitoring of blood glucose, and foot care had improved significantly in the treatment group, while no change was reported in the control group.

To make sure that the groups did not already differ prior to the intervention, they were compared on sociodemographic as well as clinical characteristics at pre-test. While results suggest that the mHealth intervention was successful in improving patient self-care of diabetes, it is important to be mindful of the possibility that those who self-enrolled into the program might already be more motivated to improve their condition.

5.2.3 BETWEEN-PERSON EXPERIMENTAL DESIGNS

Randomized Controlled Trial

The between-person randomized controlled trial (or RCT for short) is very likely the most commonly used experimental design in research involving humans. The basic logic of the RCT comes from an assumption initially made in the early 1900's within agricultural research and the invention of the concept of "random assignment." In short, the assumption is that if there are a "large enough" number of persons available to be selected from (in the classic case, fields of a crop was the "subject"), randomly assigning (i.e., like flipping a coin) these different persons to one condition (e.g., fertilizer A) or another (e.g., fertilizer B), will effectively balance out any differences that might have otherwise existed between the two comparison groups beforehand. Put differently, while factors such as soil composition, daily rain exposure, or amount of sunshine the field receives per day could likely impact growing rates, if a large pool of fields are selected from and then randomized, those differences between fields will balance out, on average, between the two assignment conditions. As such, there should be, effectively, no differences between the groups prior to random assignment because, with enough persons under study (a critical assumption that we will return to), any possible differences will eventually balance out probabilistically. This "balancing out" is the fundamental assumption behind random assignment and thus a randomized controlled trial.

Based on this, the core research question that a randomized controlled trial asks is, "does this system package result in a significant change in a target outcome relative to a control condition?" The probabilistic assumption that any differences between groups will eventually balance out with the use of random assignment provides a strong and elegant strategy for controlling for third variable factors (e.g., differences between groups that might have been present prior to random assignment that are the causal mechanism of measured differences) and thus provides a strong mechanism of control. This experimental design should be used whenever the research question is truly about an entire package (e.g., the UbitFit system that included a variety of features like a passive self-tracking mechanism, and active self-monitoring component, a live wallpaper on the phone that was a visual representation of activity, goal-setting, and other features), such as when the goal is quality assurance that some system is doing what it is intended to do. Beyond this, a randomized controlled trial can only be used if random assignment to different conditions is possible (e.g., it is not possible to randomize individuals to self-selected classes) and enough resources, particularly persons, are available to conduct the work.

While there are many advantages to the randomized controlled trial, a few notes of caution. First, the essential assumption that backs up a randomized controlled trial (i.e., that probabilistically, any group differences will eventually balance out), often requires fairly large samples to actually achieve, particularly when there are many factors that the researcher may be attempting to balance out (e.g., age, gender, race, level of technological savviness, baseline motivations to enact the change). As such, a randomized controlled trial is often a highly resource intensive endeavor to do correctly, with increasingly more work questioning if the study design should be used with smaller studies, particularly with null hypothesis significance testing statistics (Kay et al., 2016). Second, while random assignment is an elegant strategy for controlling for "third variable" factors, the causal inference that is gleaned from the study is still greatly impacted by the selection of the control group. For example, a meta-analysis focused on interventions to treat depression found that the selection of the control group (e.g., waitlist control,[30] "usual care") greatly impacted the estimate of the impact of the intervention being tested (Mohr et al., 2014). This is pivotal to highlight because the control group ultimately impacts what conclusions can be drawn. For example, if a system focused on promoting walking (such as UbiFit) is compared to waitlist control, the waitlist control condition itself might impact a person negatively by setting up the expectation that they should not start being active until they later receive the system (Mohr et al., 2014). Based on this possibility, the "control" could feasibly create a bias in your conclusions by actually stymying any natural changes that would have otherwise occurred (i.e., the core goal of the control group). As such, the same three questions highlighted with the between-person quasi-experimental design, must also be asked here.

1. What is my research question?

2. What exactly do I need to be controlling to answer my research question? and

3. What am I controlling for when I compare these groups?

Third, there is increasing interest in understanding if change is occurring on a case-by-case basis. This type of causal inference is not possible with a between-person randomized controlled trial. A randomized controlled trial provides insights about the impact of a system to incite change on average. Take, for example, weight loss trials. A very common response pattern within weight loss trials looks something like the graph below in terms of responses. In this intervention, "on average" the intervention produced more weight loss than control. As can be seen with a more careful analysis of the individuals, there are often individuals in the intervention group that the intervention did NOT work for. This type of information is lost in most analyses of randomized controlled trials. Further, even when secondary analyses are conducted to examine these individual differences in responses, it falsifies the starting assumption of differences remaining constant between groups as it is those individual differences that matter for the secondary analyses but voids the assumption of the randomized controlled trial.

---

[30]   A waitlist control involves participants being randomized to being on a waitlist until the other group has completed the trial. After that waiting period, the group can then take part in the intervention.

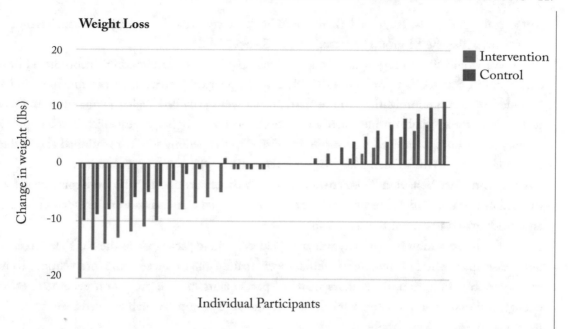

Figure 5.2: Fictitious weight loss data from a randomized controlled trial.

We highlight these points to establish that a between-person randomized controlled trial is a very valuable experimental design, but it is not without its risks. With regard to statistical analyses, a basic randomized controlled trial can often be analyzed by an advanced beginner of statistics, assuming the design is not particularly complicated (e.g., two conditions), with some statistical consultation from an expert. Further, there are a few examples in the HCI literature emerging on how to examine this type of design using Bayesian analyses that we have conducted (Lee et al., 2017). There are tomes of articles about this particular experimental design. We point you again to Shadish, Cook, and Campbell (2002) as well as several other articles that discuss factors to think about when designing a randomized controlled trial.

For example, we used this design in the pilot-testing of an adaptive physical activity intervention for overweight adults (Adams et al., 2013). The intervention consisted of a pedometer, email and SMS communication, brief health information and bi-weekly motivational prompts, along with step count goals. The participants were randomized to either the "adaptive intervention (AI)" group or the "static intervention (SI)" group. The AI group received daily step count goals that were adjusted based on the participants' own step-count on the previous days, along with micro- incentives for goal attainment. The SI group received a fixed goal of 10,000 steps per day with incentives to upload their pedometer data. The results showed that the AI group showed significantly higher improvements from average baseline steps/day to average treatment phase steps/

day as compared to the SI group (AI increased by 2,728 steps/day on average between baseline and treatment, while the SI group increased by 1,598 steps/day).

We would like to highlight a couple of methodological considerations incorporated by our colleagues in this RCT. In order to establish baseline physical activity, a 10-day run-in period was included before randomization, where participants were provided sealed pedometers and asked to continue their usual routine. Such a run-in period not only helped establish baseline physical activity levels but also helped reduce the "reactivity" to the pedometer. As mentioned above, this is particularly important in technology-based interventions where mere exposure to a novel technology may influence behavior. It also ensured that only those participants who were comfortable with the pedometer and had the technical skills to upload the pedometer data were included in the study and randomized to one of the conditions.

The intervention in the study was designed as a whole package, and the RCT design cannot help tease apart whether the observed effects were mainly due to the adaptive intervention. To help overcome this limitation, the authors matched participants in both groups on pedometer activity, educational materials, incentive amounts and on message prompts, which enabled them to be eliminated as alternative explanations.

A/B Testing

Often, you might be thinking about changing a design, but you want to see if it will solve a particular issue before making a system-wide change. A/B testing is a type of between-persons randomized controlled experiment that allows you to run small, randomized experiments on segments of your user population to see the effects of certain design changes. Since it is a randomized controlled trial, all of the information highlighted in the last section are still relevant here. With that said, within the tech industry, A/B testing is applied a bit differently and those differences are important. In brief, A/B testing is often used to help better understand smaller-scale design decisions within a much larger system. Since it is scoped down like this, it is less suspect to the issues raised above, particularly about enabling testing of a whole package compared to another package. In addition, because the differences are being implemented at a large scale, often the assumptions on balancing out between randomly assigned groups is appropriate.

Many aspects of a design can benefit from an A/B test. Perhaps you want to try an alternate onboarding flow for users that highlights specific features of the application. You can then measure how that change affects the use of those features while also measuring overall retention and usage to ensure that you have not decreased key metrics by making the change. You might also want to test smaller design changes that call attention to specific buttons or interactions in particular ways. Google reported that changing the shade of blue of advertising links led to $200M in increased revenue (Hern, 2014). "Small" design decisions can make a large impact when operating at scale, and

A/B testing allows you to learn which aspects of your design lead to specific types of use without needing to roll out a new design to all users (and potentially seeing catastrophic falls in usage for designs that do not work as hoped).

When running an A/B test, it is important to ensure that users in both buckets are as similar as possible. For example, grouping based on the millisecond that they requested a page, or if they have an even or odd user ID could be decent ways to bucket. However, bucketing based on a raw, monotonically increasing user ID would not be as good, since users with older or newer accounts might use the service very differently. It is generally a good practice to check any demographic or previous usage data for users in each of your groups to ensure that there is no statistical difference in use of the system before the experiment begins.

While A/B testing can show specific impacts to use or revenue for particular design changes, it cannot often explain "why" users are making the choices they are in the system. In the example above, the test said nothing about why users preferred to click on the slightly purple-blue link over the other colors, just that they did.

Like in-app instrumentation, A/B testing can be combined with more qualitative methods to help understand specific aspects of use, where people can articulate them. For example, you might combine an A/B test of a new version of an app or onboarding flow with an in-lab usability evaluation to have participants think aloud while walking through the new system. This might help you to understand changing perceptions of the application or its feature set that can lead to new designs and future A/B tests. Though not mobile user research per se, we have used A/B testing in combination with a variety of other qualitative and quantitative methods to improve the security warnings that users see in the Chrome browser (Felt et al., 2016, 2015, 2014; Almuhimedi et al., 2014).

However, some types of data do not lend themselves to uncovering underlying explanations. Many user interactions are subconscious and particular colors or placements of items lead to people being totally blind to links, certain types of ads, or content that you place on particular parts of a screen. Usability studies that employ eye tracking can help to uncover where people are looking, but it can be difficult to understand why, and if asked, participants might try to make up an explanation for what was not an entirely conscious decision. Sometimes, A/B tests end without a satisfying explanation for the user behavior that was observed.

Factorial/Fractional Factorial Designs: Particularly as Used in MOST

The factorial study design is a logical extension of the randomized controlled trial but with more factors and/or conditions. Specifically, a classic randomized controlled trial with one intervention arm and one control condition can also be labeled a two-factor design. A factorial design expands that logic by allowing more factors and more conditions. Returning to the UbiFit example (Consolvo et al., 2008b, 2014), the system included a variety of features like a passive self-tracking mechanism, and active self-monitoring component, a live wallpaper on the phone that was a visual representation of activity, goal-setting, and other features. In a classic randomized trial, all of those components would be tested as a package relative to control. In a factorial design, all of those components would be separated and recombined within the study to examine their individual and synergistic effects. So UbiFit could have feasibly been broken up into a factorial design to have the self-monitoring component or not, the live wallpaper feature or not, and the goal-setting feature or not, while maintaining some common elements that would always be present such as passive self-monitoring. This would be represented as a 2x2x2 factorial design (see below).

Table 5.2: Plausible factorial design to test some of the UbiFit system's individual components

Experimental Conditions	Factors		
	Self-monitoring	Live Wallpaper	Goal-setting
1	x	x	x
2	x	x	✓
3	x	✓	x
4	x	✓	✓
5	✓	x	x
6	✓	x	✓
7	✓	✓	x
8	✓	✓	✓

While factorial designs are not new, and indeed, conceptually the same thing as a randomized controlled trial but with more factors, the work of Dr. Linda Collins and others has been advancing strategies for using the factorial and fractional factorial design as a method for developing optimized interventions in a highly resource-efficient manner. A full discussion on the subtleties about how exactly this resource efficiency is achieved is beyond the scope of this write-up but we point you to resources on this, notably, *The Methodology Center at Penn State*.[31] Some key highlights and advantages of the factorial/fractional factorial design as used in the Multiphase Optimization Strategy (MOST) is that multiple research questions can be answered simultaneously and effectively in a single study that often requires FEWER resources than running multiple randomized controlled trials for each component (Collins et al., 2007). Further, the study design allows for a test of if there are interactive effects between intervention components, which means that causal inference can be gathered about whether an intervention component only "works" when some other intervention component is present (or perhaps even the possibility that some components do NOT

[31]   See https://methodology.psu.edu/ {link verified Dec 28, 2016}.

work, when presented together as they interact antagonistically). Overall, the emphasis on perpetual optimization (an essential emphasis of the MOST strategy), coupled with a focus on components, and the testing of components in a resource-efficient manner make this experimental design very well-suited for mobile user research endeavors. We strongly recommend careful consideration on the use of this experimental design, in particular, as it is often well-matched to the implicit questions mobile user researchers would be interested in asking, but did not previously believe they could because it would be too resource-intensive.

In terms of the factorial/fractional factorial design as used in MOST, the core research question explored is, "What is the independent impact of each component of a system on a target outcome?" A secondary question that can also be answered is, "What is the interactive effect of each system component on a target outcome?" Just like a randomized controlled trial, the use of random assignment with enough samples provides an elegant strategy for controlling for third variable factors that might drive cause and effect relationships. Unique to this study design is the possibility of testing multiple research questions about each component simultaneously in a highly resource-efficient manner and the ability to glean causal inference about the interactive impact of system components (Collins et al., 2005). Further, this study design is also less susceptible to the issues of selecting an appropriate control condition because the design implicitly builds in meaningful levels of controls by only providing or not providing different intervention components. As such, causal inferences are more localized (i.e., at the "molecular" rather than "molar" level, discussed below) and thus can be more easily linked to the specific components. This is in contrast to the randomized controlled trial, which classically in behavioral science work, had multiple components varying simultaneously between intervention and control (and, as an aside, the randomized controlled trial, as a methodology only suffers from this issue because of the selection of the control group and the clarity of the research question, not because of the methodology itself). A key disadvantage of this study design is that causal inference is still occurring at the between-person/average level. As such, insights on change that might be occurring on a case-by-case basis are not well supported. A second disadvantage of this experimental design is that to conduct it effectively requires a fair degree of training, both in the analyses and the study design. That said, the methodology center at Penn State, led by Collins, has an impressive array of training materials online and also regularly runs training sessions to teach individuals about this process. As such, there are great resources to help enable mobile user researchers in the use of this method.

An example of the use of a fractional factorial design as part of the MOST framework is optimization of a web-based smoking cessation intervention by Strecher et al. (2008). This work aimed to identify the active psychosocial and communication components of the program and examine the impact of increased tailoring on cessation. This study examined five intervention components: outcome expectations, efficacy expectations, use of hypothetical success stories, personalizing the message source, and the timing of the exposure to messages. Each of these factors had two levels; every participant received one variation of each (see Table 5.3).

Group	Exposure	Outcome expectation depth	Success story depth	Efficacy expectation depth	Source personalization
1	Multiple	Low	Low	Low	High
2	Multiple	Low	Low	High	High
3	Multiple	Low	High	Low	Low
4	Multiple	Low	High	High	Low
5	Multiple	High	Low	Low	Low
6	Multiple	High	Low	High	Low
7	Multiple	High	High	Low	High
8	Multiple	High	High	High	High
9	Single	Low	Low	Low	Low
10	Single	Low	Low	High	Low
11	Single	Low	High	Low	High
12	Single	Low	High	High	High
13	Single	High	Low	Low	High
14	Single	High	Low	High	High
15	Single	High	High	Low	Low
16	Single	High	High	High	Low

Table 5.3: Experimental groups of the 16-arm fractional factorial design (Strecher et al., 2008)

"Single" exposure refers to the smoking cessation content provided in one large set at one time, while "multiple" refers to providing it as a series on a weekly basis. "High" and "Low" outcome expectation depths refer to the high or low levels of tailoring and individualization done in providing feedback and advice on the expectations participants have from quitting, and the motives for quitting. "High" and "Low" success story depths refer to the depth and extent of tailoring of the hypothetical success stories that the participants received as part of the intervention. Personalization of source refers to the introductory session of the intervention, where "high" personalization included a highly personalized welcome to the participant and "low" personalization used an impersonal version. Efficacy expectation depth refers to the tailored efficacy messages that were sent to the participants that addressed the barriers to cessation. "High" depth included highly tailored feedback to the individual's two most important barriers, while "low" depth included less-tailored content addressing broad areas of barriers.

The use of a fractional factorial design in this study allowed the researchers to observe the main effects of all five components along with several two-factor interactions. The findings suggested that smoking abstinence was most influenced by high-depth tailored success stories and high source personalization, and that a cumulative assignment of all tailoring depth factors resulted in higher rates of six-month cessation (Strecher et al., 2008).

Sequential Multiple Assignment Randomized Trial (SMART)

The Sequential Multiple Assignment Randomized Trial (SMART) (Murphy, 2005) is an experimental design developed by Susan Murphy, Linda Collins, and others as part of The Methodology Center at Penn State. Conceptually, it is a different type of "optimization" method that fits within the "MOST" strategy, just like the factorial/fractional factorial. The difference, however, is that SMART allows for a different research question to be answered. Unique to the SMART trial is that it provides causal insights, not about the system package or components, but actually about the decisions that are made based on responses to a system. A classic example from a medical research context is answering the question, "what do I do if someone is not responding to the intervention?" Take, for example, a weight loss trial. What should be done if a person has not lost any weight after three months with the intervention? The SMART trial provides an empirical strategy to test whether the decision rule developed for adapting an intervention is useful or not.

A full discussion on exactly how this study design achieves this is well beyond the scope of this brief summary. Important for mobile user researchers is the awareness that such an experimental design exists to test adaptation. As the study design uses random assignment, it has the same advantages and disadvantages of a randomized controlled trial (e.g., good control for third variable factors as the between-person level). Unique to a SMART trial is that it can control for the "main effect" impact of an intervention or its components and then parse out the unique impact of any decisions made depending on the response of individuals in the trial. The key disadvantages of this study design are that it is a very complicated study design that requires advanced statistical knowledge and advanced knowledge on how to properly design the trial. Further, similar to all other between-person designs, the randomization—while taking into account individual responses —is still testing the decision rule for adapting on average. As such, limited within-person causal insights can be gleaned. Similar to the factorial/fractional factorial design as used in MOST, The Methodology Center at Penn State, as well as Susan Murphy's research group at the University of Michigan, particularly Daniel Almirall, provide extensive training on the use of this method, including providing direct consultation to researchers. There is also a great deal of resources online to get started in the use of this design.[32]

---

[32] See https://methodology.psu.edu/ {link verified Dec 28, 2016}

Unfortunately, this study design is not yet common in mobile user research and thus we will provide an example outside of the mobile user research realm and then briefly discuss possibilities for mobile user research after this example. An example of the use of a SMART design is the Reinforcement-Based Treatment for Pregnant Drug Abusers (RBT) study (Jones 2010; Lei et al., 2012). RBT is a multicomponent intervention that consists of elements that focus on non-drug related reinforcers (i.e., factors that support the continuation of a behavior) to reduce drug use or changes of relapse, such as behavioral contracts, life skills training, reinforcers contingent on abstinence (e.g., providing a reward when a person does not drink), functional assessment of drug/alcohol abuse, social clubs, etc. The rationale behind the study is that although RBT has been shown as an efficacious intervention, it is time consuming for the participant, and about 40% of the participants do not respond as desired. The SMART trial was designed to answer questions such as whether the treatment-as-usual approach for RBT can be reduced in intensity and scope, whether someone who does not respond to the treatment-as-usual approach should be moved to a more intensive version or continue on the same, and whether the intensity can be reduced for someone who responds quickly to the first approach. In this particular design, four variations of the multi-component RBT were considered and tested in various adaptive interventions. These variations, in the order of intensity and scope, were aRBT (least intensive), rRBT, tRBT (treatment-as-usual), and eRBT (most intensive).

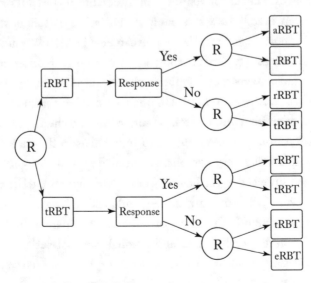

Figure 5.3: Experimental design of a SMART trial. From Lei et al. (2012).

All participants were initially randomized to either rRBT or tRBT, and response to treatment was assessed at two weeks. Depending on the response, individuals were further randomized to specific treatments based on the tailoring rationale. An example of one of the adaptive interventions was to begin treatment with tRBT and stick to tRBT if the patient showed an adequate response at two weeks, otherwise, switch to eRBT (Lei et al., 2012). Although not from the technology realm, this study provides a good example of the scope of SMART designs as it provides insights on the sorts of questions that can be answered. It can explore questions about what to do if individuals do not respond to an intervention or what to do if a person responds very quickly to an intervention. These are all possibilities within mobile user research, thus highlighting the allure of this potential design.

5.2.4 WITHIN-PERSON EXPERIMENTAL DESIGNS

Micro-randomization Study

The micro-randomization study design is still being developed as a methodology, but it holds great potential for mobile user researchers, in particular, and thus we wanted to highlight it here. The micro-randomization design could also be labeled a "sequential factorial design" in that the general idea of the design is to randomize receiving or not receiving different intervention components over time (sequentially) on a case-by-case (within-person) basis. For example, you could easily take the general factorial design we proposed above for testing the components of UbiFit (i.e., active self-monitoring, goal-setting, and the live wallpaper) and feasibly do the randomization within-in-person by randomly assigning different days to have a person self-monitor or not or set goals or not. For the live wallpaper, it might seem strange to the user to have it present and then absent one day to the next; as such, expectations of the user are important to take into account when designing a micro-randomized trial. For example, rather than randomizing if the live wallpaper will be present or absent each day, the study could randomize each week or even each month to having the live wallpaper or not. It is not a problem to have multiple randomization signals working at different timescales within a micro-randomized trial. Indeed, it is plausible to randomly assign certain components within a day to whatever timescale could be meaningful, such as every few hours (more on this with the concrete example given below). Likely most attractive for mobile user researchers is that a micro-randomized trial is a highly resource-efficient strategy for testing out a mobile system because there are repeated exposures occurring on a case-by-case basis. This allows both for within-person/single-case causal inferences to be feasibly gleaned (although one needs to be very careful with the statistics) and also allows for those insights to be gathered with relatively few subjects, usually in the realm of participants often used in mobile user research-related field studies (e.g., in the ballpark of 25-50 participants). As such, it is an excellent design to be considered for mobile user research.

The key research question that a micro-randomized trial is optimized to answer is: "What are the short-term (proximal) effects of an intervention component on impacting a target proximal outcome behavior?" Beyond this, a micro-randomized trial can also examine secondary questions including: (a) How do the short- and long-term effects of an intervention component change over time? (e.g., in a system like UbiFit, goal-setting might only be useful at the beginning of an intervention whereas the live wallpaper might be particularly useful later in the intervention); and (b) Which factors, both fixed and time varying, might moderate the effect of an intervention component over time? (e.g., on days when a person is more stressed, perhaps the goal-setting intervention will be less effective).

This experimental design can be used whenever (1) an intervention component can be feasibly activated and deactivated without greatly impacting the intervention, and (2) the expected effect of a system component is perceived to occur relatively quickly (e.g., an intervention to increase steps is thought to primarily impact that day's step goals). As such, this experimental design is not appropriate if the impact of the system component is believed to have a very slow response (e.g., it takes in the realm of weeks to months before an effect can be observed). The key advantages of this design are that it is an elegant experimental design that, via random assignment, provides a robust mechanism for controlling for "third variable" explanations, uniquely, at the within-person level. As such, it is a good experimental design for testing out the impact of a system over time on individuals. It is also an advantageous design because it can be conducted efficiently with relatively few participants and thus fits with current resource constraints often placed on mobile user researchers for field studies. The key disadvantage to this experimental design is that it is a complicated design that requires both an advanced understanding of experimental designs and advanced knowledge in statistics to analyze the data. Further, since it is still being developed, there are few training programs or other options available for learning how to develop this sort of experimental design. That said, Susan Murphy at the University of Michigan is increasingly developing the methods, processes, and papers to articulate how to enable others to use these methods within their own work and, similar to the MOST factorial design or SMART trial, it is likely that within the next few years, there will be a wealth of training opportunities for researchers to learn how to use this method. As such, in a few years, this will likely become a very valuable and accessible methodology for mobile user researchers.

An example of this is our study, HeartSteps (Klasnja et al. 2015). HeartSteps consists of an Android application intended to encourage walking and a wrist-worn fitness tracker that measures the user's steps taken throughout the day. The intervention is made up of two components: daily activity planning and contextually relevant suggestions for physical activity. The daily activity planning is done on the previous evening and involves planning and specification of when, where and how the user will engage in the behaviors that improve physical activity. The contextually relevant suggestions are of two types: (a) to get up and go for a walk and (b) to stop being sedentary (i.e., get up and move). Participants are randomized daily to either daily planning or no planning, and also to receive contextually relevant suggestions or not. Diving further into the contextually relevant suggestions, there are five potential times of day when the suggestions could be delivered (morning commute, lunchtime, mid-afternoon, evening commute, and dinner). In this design, to reduce burden from repeated notifications from the app, the authors chose to give two activity suggestions daily, and whether the participant receives the suggestions or not is also randomized to any two of the five available times per day. Such a design can help investigate the proximal or direct effects of the intervention components (e.g., the user walking right after they receive a suggestion, and effects of the suggestion measured in terms of number of steps within 60 min after receiving the prompt, or the number of steps on the day following the suggestion for daily planning) along with the distal effects (reach and sustain recommended levels of physical activity over time). It should be kept in mind that an important concept in micro-randomization trials is that the intervention delivery is contingent on the user availability and appropriateness of intervening at that particular moment. Going back to HeartSteps, this would mean that the participant would not be prompted at inappropriate times (e.g., if they were already walking or driving).

5.2.5   OTHER DESIGNS

The above experimental designs represent only a small portion of plausible experimental designs that could be used for mobile user research and, indeed, there are other experimental designs that are being developed that better match a mobile user research context. For example, David Mohr and colleagues have been developing a methodology they have labeled the continuous evaluation of an evolving behavioral intervention technology (CEE-BIT) (Mohr et al., 2013). In CEE-BIT, the basic logic is to use older versions of a system as the "control" for newer system releases. As such, it is an elegant strategy for fostering optimization that fits into current development strategies and thus has a high potential for future mobile user research.

We, along with Rivera and other colleagues, have been further advancing the logic of a micro-randomized trial by incorporating lessons from control systems engineering principles into the design of the randomization signal. The key advantages to this design are that it will enable far more subtle testing and exploration of the various time-effects that could occur on a case-by-case basis and the ability to study continuous, as opposed to simply categorical, interventions. For example, one system ID experiment for mHealth interventions that we designed was developed to create dynamical models for defining a daily "ambitious but doable" daily step goal (Martin et al. 2015, 2016). This target of an ambitious but doable daily goal is a deceptively hard target as myriad factors can conceivably impact what is "ambitious but doable" for any given day such as previous behavior, mood, weather, or the interactions with others. In addition, it is important to note that this experimental design, when used as part of system identification, has a different purpose. The goal is to create a dynamical model for mathematically describing a phenomenon, which can then be incorporated into a controller (for the interested reader, see Martin et al., 2015, 2016). While much still needs to be examined about this design, we wanted to highlight it as a potentially viable option for future work.

5.2.6   GENERAL WORDS OF CAUTION

It is important to briefly discuss the competing interests that often go into choosing an appropriate experimental design. As discussed earlier, the goal is generalized causal inference. Classically, a distinction is made between the "internal validity" compared to the "external validity" of a study. *Internal validity* (also labeled *local molar causal inference*) refers to the overall confidence that a causal inference (e.g., this intervention impacts this outcome) is correct. It is "local" in that internal validity does not extend beyond the specific elements of the study (i.e., the specific intervention/ system, the specific measures used to assess the factor you wished to change, the specific individuals who took part in the study, and the particular setting whereby the study took place), and it is "molar" because the causal inference is often a package of components (as opposed to a "molecular" causal inference that would be about the smallest meaningful components). *External validity*, on the other hand, establishes the extent to a which any causal inference can be generalized across different intervention/system options (e.g., since goal-setting option 1 worked, goal-setting option 2 should also work), observation (e.g., since it changed self-report physical activity, it should also impact objectively measured physical activity), individuals (e.g., since it worked for the specific 25–65 year old women in this study, it should work for all 25–65 year old women), and settings (e.g., since it worked in this neighborhood in Phoenix, it should also work for a neighborhood in New York City). Unfortunately, strategies for improving internal validity (e.g., restricted range of individuals to be studied via stringent inclusion/exclusion criteria, conducting the study in controlled settings such as within labs) can be "threats" to external validity whereas strategies for improving external validity (i.e., randomly selecting individuals to use from an entire population) often can be "threats" to internal validity. This compromise provides the logical foundation for the distinction between two types of randomized controlled trials that you might have heard of: the efficacy trial and effectiveness trial. Efficacy trials are meant to establish causal inference (internal validity) within "ideal settings," and effectiveness trials are meant to establish causal inference across individuals and settings (external validity).

As such, one word of caution is that, often, a study must balance these two competing desires within research (i.e., good causal inference and good generalizability so that the insights can be applied to novel circumstances beyond those studied). Therefore, it is important for any mobile user researcher to be clear on whether their desire is to have a good sense of what happened specifically in the target group studied vs. something that is generally true across different individuals and contexts. For mobile user research, there is often a stronger desire for externally valid insights in "the real-world" (a core reason to go mobile). While this does present challenges to inferring if a mobile system "worked," we will show how there are experimental designs that do match this purpose very well, assuming the research question is more focused on the components rather than an entire system. As such, we are advocates of factorial designs and micro-randomized trials as particularly valuable for mobile user researchers (see Klasnja et al., 2017 for a more detailed discussion on this).

Beyond this, it is important to acknowledge two other issues that impact all experimental designs:

1. what you use as your "control" greatly impacts the conclusions you can draw, and

2. who specifically was in your study greatly impacts how likely your results will be true elsewhere.

The issue of the impact of the "control group" is discussed in great detail by Mohr et al. (2014). For example, if you were interested in knowing the impact of an intervention to increase worksite productivity, you could feasibly "control" for this effect by comparing individuals who received it to those who were measured without an intervention. This "control" strategy, however, means that there are many factors (e.g., likely more contact time, your entire intervention, possible interactions that occur because of the intervention but were not intended) that differ between the two. As such, at the end of your intervention, you might be able to say that "it worked," but it will be hard to know "how," as knowing exactly what worked is still a mystery.

On the second point, it is critical to remain mindful of the impact of the specific people with whom you test an idea. For example, if they are "WEIRD" (White, Educated, Industrial, Rich, and Democratic), your results might not generalize outside of the WEIRD group (Brookshire, 2013). In general, choosing participants from university settings is not recommended as they do not represent the larger population (e.g., in the U.S., two-thirds of the adult population does not have a college degree). This is why taking the time to clearly define a target user group is so essential, as it greatly impacts what conclusions can reasonably be drawn from an experiment, particularly if there is a desire to have "lessons learned" that could translate to a new context (e.g., a point we also mention when selecting a good theoretical model in Chapter 6).

5.3   SUMMARY

In this chapter, we highlighted many different experimental designs you can employ to answer the question, "does it work?" Our goal was not to teach you how to conduct all of these designs but, instead, to provide you with just enough information about the various experimental designs to help inform your selection of an appropriate design for your mobile user research question, when the goal of your research is to elicit a change. As we highlighted, there is a wealth of valuable experimental designs, each of which have strengths and limitations. Ultimately, the selection of an experimental design should be dictated by a clear understanding of the research question you are attempting to explore. We hope that you will use this chapter to help you not only select an appropriate experimental design, but also expand your perception on the possibilities for testing exactly what "it" is that "works" within a mobile user research study.

We particularly recommend a careful examination of the factorial/fractional factorial design as used in the MOST framework and the SMART trial, as well-established designs. We also recommend the micro-randomization study and CEE-BIT designs as emerging experimental designs for mobile user researchers, as they are well aligned with the types of research questions often implicitly explored by mobile user researchers, while also being well suited for current resource constraints.

CHAPTER 6

Using Theory in Mobile User Research

"Experience without theory is blind, but theory without experience is mere intellectual play."

Immanuel Kant

6.1 INTRODUCTION

Mobile user researchers may have heard suggestions related to using theory within their work. A good theory can indeed help provide a scaffolding around any mobile user research endeavor. But what is a theory? How can a theory help you perform more effectively? How can you "use" theory to better understand potential users? The goal of this chapter is to provide answers to these questions by providing pragmatic advice on how to interpret and use theories within mobile user research.

While there are many possible uses of theory, in this chapter, we will emphasize three. The first use is to provide a structure for developing and choosing appropriate measures, including the design of semi-structured interviews or observational studies as well as more quantitative measures such as self-report scales. The second use of theory is to better specify and define the target audience for any mobile user research endeavor. Finally, the third and often dominant use of theory is to help guide the design and overall development of technical artifacts, such as smartphone apps, to promote behavior change or self-tracking strategies.

Prior to these uses though, we start by first providing a list of terms that we used in Hekler et al. (2013a), as these terms will be used throughout the chapter. We then turn to a more in-depth discussion of the three major uses of theory within mobile user research. This is followed by a discussion on "words of caution" with the use of current behavioral theories as the theories are "works in progress" and thus have important potential flaws to be mindful of. We then discuss strategies for picking the "right" theory for your area of inquiry, particularly if you have limited knowledge of theories. We follow that with a few prominent theories to get you started and then discuss a case-study, the MILES study, which utilized theory extensively in the design and evaluation process. We conclude with a summary of key points from the chapter.

6.2   DEFINING TERMS

Glanz and Rimer (1995, p. 4) define theory as "*...a systematic way of understanding events or situations. It is a set of concepts, definitions, and propositions that explain or predict these events or situations by illustrating the relationships between variables.*" In this chapter, when we mention theories, we will be using this definition. Put differently, theory refers to any organizing structure for understanding behavior or other targeted phenomenon. There are important "building blocks" of theory that also require defining, including:

- **constructs**, which are the fundamental components or "building blocks" of a theory, (e.g., two constructs from the Technology Acceptance Model are *perceived usefulness* and *perceived ease of use* of a technology) (Davis, 1989)and

- **variables**, which are the operational definitions of the constructs, particularly as they are defined in context (e.g., specific measures used to assess perceived usefulness of a technology, or strategies used within an application to influence the perceived usefulness of a technology).

There are many ways in which theories can be classified, for example, based on whether they are **descriptive theories** that explain the determinants of human behavior, such as the *Theory of Planned Behavior* (Ajzen, 1991), or those that conceptualize the **process of change**, such as the *Transtheoretical Model* (Prochaska and DiClemente, 1994). For the purposes of this chapter, we use the classification that we've used previously (see Hekler et al., 2013a), and categorize theories based on their level of generality/specificity, as follows.

Conceptual Frameworks focus on a specific facet of a problem. The advantage of these theories are that they tend to provide more specifics for understanding and studying a particular domain. Examples of conceptual frameworks are *goal setting theory* (Locke and Latham, 2002), *self-efficacy theory* (Bandura, 1977), and so on. Goal-setting theory, for example, has been used in a variety of HCI systems, such as our UbiFit system (Consolvo et al., 2009). Goal-setting theory provides specific insights on strategies for setting actionable and appropriate goals.

Meta-models are broad organizational structures for understanding behavior. Theories that function more as meta-models are short on specifics but very valuable in terms of defining the landscape and broad domains of inquiry, as they are often true across a wide range of behaviors. Put differently, they are valuable for setting up a broad perspective for starting to think through a problem, but many details will still need to be filled in via conceptual frameworks and other mobile user research methods such as *experience sampling* (covered in Chapter 4).

An example of a meta-model which is popular in HCI is *Fogg's Behavioral Model* (Fogg, 2009). This model proposes that for a target behavior to occur, a person needs to be sufficiently *motivated*, possess the *ability* to perform the behavior, and be *triggered* at an opportune moment to perform the behavior. While Fogg goes into greater detail on each, the high-arching structure that emphasizes motivation, ability, and triggers, helps identify the different levels of inquiry possible in relation to a particular domain (Hekler et al., 2013a).

Unfortunately, most theories are not labeled with these terms. While it might seem pedantic, this distinction is valuable to be mindful of when trying to engage with theories, as it helps to determine how best to use the theory. Meta-models and conceptual frameworks are on a continuum with regard to specificity and generalizability. As such, an important first step is to think about how generalizable or specific the theory you're considering is. If it is vague and likely to be true in general, it is more likely a meta-model and thus useful for providing a rough scaffolding for additional inquiry. If it is very specific and detail-rich, it is more likely on the side of a conceptual framework and thus will be more useful for specific design decisions, but may not be a good match for your particular study or system. We will return to this distinction in theories throughout the chapter.

6.3 USES OF BEHAVIORAL THEORY

At a high level, a theory provides a reference point that can help you better understand a problem or devise new and innovative solutions to a given problem. As such, a theory provides you with a sort of "reference point" you can use to judge your work and to provide a structure for organizing a mobile user research project. There are, of course, other reference points, particularly previous work and the target user. Previous work (e.g., the scientific literature or previous technical systems developed) helps to determine the current "state of the art/science" for a given problem, which can be helpful for defining an important problem to work on, building on current best practices, and avoiding the possibility of "reinventing the wheel." Target users, as a reference point, help to provide the rich and subtle complexities that are particularly true for the problem you are tackling (and is a key focus of other chapters). We do not discount the importance of these reference points and, instead, see the three as complementary reference points that should be utilized throughout any mobile user research project. Theory as a reference point is particularly valuable for three types of mobile user research activities:

1. understanding the target problem and devising better ways to observe, measure, and study it;

2. defining a target audience; and

3. defining the design of any technical system.

Next, we discuss each in turn.

6.3.1   UNDERSTANDING THE TARGET PROBLEM: DESIGNING WAYS TO OBSERVE, MEASURE, AND STUDY

Assuming you have chosen a generic target problem (e.g., encouraging people to be physically active) and a target group (e.g., older adults), theories can help in providing a structure to understand and define the problem in more depth. The distinction between meta-models and conceptual frameworks is particularly important when attempting to better understand the target problem and user. As such, we will describe their uses separately.

Using Meta-models to Guide Measurement

Meta-models can be useful for defining measures, but at a broad range of inquiry. For example, the *COM-B Model* (Michie et al., 2011) (which is similar to Fogg's *Behavioral Model*) postulates that for any behavior to occur, there must be the appropriate balance of a person's **c**apability to engage in the behavior, **o**pportunity within a person's context to engage in the behavior, and **m**otivation to engage in the **b**ehavior (hence *COM-B*). This broad heuristic can be very valuable for determining what to "measure" at many stages of your work including initial formative (also called *generative* or *exploratory*, as discussed in Chapter 3) work and observation. For example, let's imagine you were interested in studying water usage with the eventual goal of building a system that fosters water conservation. The COM-B model provides a generic structure to start organizing questions for semi-structured interviews (e.g., Capability—*Tell me about what makes it easy or hard for you to conserve water*; Opportunity—*Tell me about areas in your home where you use the most water*; Motivation—*What motivates you to use water? to conserve water?*). While these broad questions might not facilitate strong insights right away, they do provide a rough skeleton for fostering inquiry and also provide a heuristic for organizing observations and findings from previous literature. To continue with the example, after conducting many hours of observation focused on water usage in a home environment, a researcher could organize observations as potential capability (e.g., limited knowledge on how to program a sprinkler system), opportunity (e.g., using inefficient shower nozzles), or motivational (e.g., water wastage is not perceived as a problem) problems to be addressed. A meta-model could also provide structuring for the development of targeted surveys, both broad-scale and more targeted experience sampling questions (e.g., Behavior—*How long was your shower today?* Opportunity—*How much time do you usually have to shower?* Motivation—*How motivated were you to conserve water while taking a shower this morning?* Capability—*What is the shortest amount of time you need to have a "good" shower?*). In summary, a meta-model is often a good framework to start areas of inquiry, particularly when very little is known about a given problem within a particular target population.

The work of Robinson et al. (2013b) provides another example of the use of COM-B for defining measures and areas of inquiry as a generic meta-model. They first used the reference point of previous work by conducting a systematic review and meta-analysis of the studies that examined the influence of awareness and memory on food intake to support the design of a smartphone app (Robinson et al., 2013a). They then used the COM-B model to describe their potential user in terms of their capabilities (individuals may lack knowledge of the potentially harmful effects of non-attentive eating), opportunities (individuals may not have the tools necessary to record their food intake and increase awareness of food consumption), and motivation (non-attentive eating may be occurring out of habit) in context of the desired target behaviors (memory recall of earlier food consumed, prior to new food intake). Using such a model provided them with a framework to categorize the various levels of influence that they believed affect attentive eating in individuals. A second example in HCI is Lee et al. (2017), in which a meta-model was used to support individuals in self-experimentation. In this context, the "designer" was the individual themself, but the same basic logic in terms of using the meta-model as a framework for thinking still holds true.

Using Conceptual Frameworks to Guide Measurement

If more details are known about a given domain, it might be possible to find acceptable conceptual frameworks to help guide work. Conceptual frameworks are quite valuable if they can provide insights that are specific to the domain of inquiry. For example, imagine you were working on doing some mobile user research focused on understanding work efficiency. Various theories from the field of industrial organizational psychology provide a myriad of frameworks for defining what to measure. For example, the *Motivation-Hygiene Theory of Job Attitudes* is a conceptual framework developed by Frederick Herzberg and colleagues (1959) that talks about the two main categories of factors that influence job motivation, satisfaction, and dissatisfaction. These are: (1) *growth/motivator* factors and (2) *job-dissatisfaction avoidance/hygiene* factors. The motivator factors are those that, when present, lead to higher motivation and job satisfaction. These factors are achievement, recognition for achievement, the work itself, responsibility, and growth or advancement. The hygiene factors are those, that when absent, lead to job dissatisfaction. These factors are company policy, technical supervision, salary, working conditions, and interpersonal relations (Herzberg, 1968). If understanding worker satisfaction is one of the goals of your user research, a theory that provides very specific insights on factors related to job satisfaction, such as Herzberg, Mausner, and Snyderman's, can help you define the measures. For example, the theory suggests the importance of asking questions related to motivators like recognition of achievement (e.g., *How often do you feel that you are recognized when you do your job well?*) or fulfillment from the work itself (e.g., *How much do you enjoy doing the everyday tasks that are required in your job?*) or related to hygiene factors such as questions related to company policy issues (e.g., *tell me about the company policies that make it enjoyable to work here. What about the ones that make it hard to work here?*) or interpersonal relations (e.g., *how collegial do you feel your workplace is?*). **In contrast to the meta-models, these provide very specific factors to be mindful of for the target problem.**

While conceptual frameworks can plausibly provide much direction, it is often the case that conceptual frameworks are not well-specified to a given problem area. For example, *self-efficacy theory* (Bandura, 1977) emphasizes, among other things, the importance of self-confidence (labeled *self-efficacy*) as a key determinant of a given behavior. This is a specific prediction but it has been applied to a wide range of domains including health behaviors (e.g., physical activity, smoking, sleep) and sustainability behaviors (e.g., water usage). When a conceptual framework provides a specific prediction but it can then be applied across a wide range of domains, it is important to be wary of this prediction. It is quite plausible that the prediction made in one of these theories is not relevant for your given problem domain.

For example, a key component of the *transtheoretical model* (Prochaska and Velicer, 1997) is the concept of the "stages of change." This theory suggests that individuals go through different phases or "stages" within any given change process. The process of change, according to the transtheoretical model, starts with being a *pre-contemplator*. Precontemplators are those who are not thinking actively about a particular problem (e.g., the need to quit smoking). The next stage is *contemplation*, where the individual is thinking about the problem, but not acting. The next stage is *preparation*, whereby an individual is actively working to plan a strategy of change. *Action* is the next stage and, as the name implies, involves a person actively attempting to change their behavior. The final stage is maintenance, which is when a person has successfully enacted a new behavior and is thus "maintaining" the behavior over 6 months or longer. These stages of change were developed particularly to understand smoking cessation, but they have been applied to a wide range of domains (Prochaska et al., 2008). This lack of specificity on when a theory is, and more importantly is not, relevant is vital to be mindful of during development work as it is quite possible to pick the wrong conceptual framework and thus be led astray. Put differently, picking the wrong conceptual framework can lead you to the wrong questions and the wrong area of inquiry, thus diminishing your ability to find the answers you seek for your mobile user research problem. For example, if you are interested in inspiring someone to get vaccinated, that is often a one-time event. As such, the concepts of the stages of change, particularly action and maintenance, may not be as important.

6.3.2   DEFINING A TARGET USER AND AUDIENCE

A theory can also be very valuable to help further define and select target users from something generic (e.g., obese individuals) into something far more actionable (e.g., obese women who are motivated to lose weight). Many theories suggest that different user groups will have different needs, and that an intervention that may be effective for one group may not be effective for another (Hekler et al., 2013a). As an example, consider the *Transtheoretical Model* (TTM), which we mentioned earlier that incorporates different stages of change (Prochaska and Velicer, 1997). This model posits that for an intervention to be effective, it needs to be matched to the stage of change the individual is in. As an example, during the development and evaluation of our UbiFit system, a mobile-based system designed to encourage individuals to increase physical activity, we recruited only those participants who wanted to increase their physical activity (contemplation, preparation, and action stages) and screened out those in the "precontemplation" and "maintenance" stages (Consolvo et al., 2008a, 2008b). These were individuals who, at that time, did not want to increase their physical activity, or who were already doing a good job of maintaining their physical activity, respectively. This screening was done under the assumption that the features of UbiFit may not be useful to a person who has no interest in being more physically active or has already figured out a strategy that works for them.

Theories can also help determine and predict the target audience of a particular technology. For example, the *Diffusion of Innovations* theory by Rogers (1962), which was popularized in "Crossing the Chasm" (Moore, 1991), describes the process whereby an innovation or new idea gets transmitted over time amongst the members of a social system. Part of this theory categorizes the users or "adopters" of a system based on their interest and acceptability of *innovativeness*, which is the degree to which an individual is likely to adopt a new idea. This framework recognizes five categories of adopters; innovators, early adopters, early majority, late majority, and the laggards. Many differences between the characteristics of these earlier and later adopters based on personality variables, socioeconomic status and communication behavior have been shown between these groups. For example, earlier adopters have a higher social status, more years of education, are more highly interconnected in the social system, and may be less dogmatic than the later adopters (Moore, 1991). Although some of these generalizations should be applied with caution, a theoretical framework such as this provides a scheme to identify target audiences (for example, the "early adopters" of a new physical activity app would be those individuals interested in self-tracking, being physically active, and have an interest in technologies for their own sake and be willing to put up with "bugs" and other features that are not yet fully fleshed out). Apart from the Diffusion of Innovations, researchers can also utilize other models such as the *Technology Acceptance Model* (Davis, 1989), and the *Unified Theory of Acceptance and Use of Technology* (Venkatesh et al., 2003) that originated in the field of computer science to get insights into the factors that affect the acceptance and adoption of new technologies. Such theories have been applied in a vast number of domains such as to understand the factors that influence acceptance of mobile banking (Luarn and Lin, 2005), to predict intentions to use online shopping (Vijayasarathy, 2004), and adoption of mobile healthcare in hospital professionals (Wu et al., 2011).

In each of these uses, we have also highlighted how the use of theory provides a reference point for judging the work being accomplished. A precondition to getting to this point is finding the "right" theory for your given problem. It is a surprisingly tricky problem to find the "right" theory and thus a core point that we will address next.

6.3.3   DEFINING THE DESIGN OF A TECHNICAL SYSTEM

The most common use of theories is to support the design of a technical system. They can be useful for defining the core features of the system, insights on user interfaces, and on the likely user experiences that would be most desired for the guiding problem. Theory can help you define what features to include, as well as the underlying purpose for the features. This can be very valuable when conducting user research, as it allows the researchers to link the feature (e.g., graded goal-setting) to a specific measure (e.g., self-efficacy) to see if their feature is impacting the desired outcome (Klasjna et al., 2011). Further, theory, particularly an awareness of different theoretical constructs, can be very valuable for making new design decisions when desired outcomes are not occurring.

There are many good examples of this type of work in the HCI literature (e.g., several were highlighted in Hekler et al. (2013a). Returning to the Robinson et al. example mentioned earlier, they used the factors identified from the COM-B meta-model to determine what features or tools may help users to engage in attentive eating (Robinson et al., 2013b). Figure 6.1 illustrates how they used the COM-B structure to inform the design of the system.

Table 1 BCW strengths of smartphone technology to help users eat more attentively

Capability	Opportunity	Motivation
✓ Smartphone technology can allow for faster recording (embedded camera and touch screen input) and relay (slideshow presentation) of information, in comparison to traditional paper based tools. This strength should make completing target behaviors easier (capability), increase the likelihood that users will have time to complete target behaviors (opportunity) and make behaviors less arduous (motivation).		
✓ Storage and relay of eating episodes in technology increases capability of achieving key target behaviors.	✓ Smartphones are widely used, which should ensure: 1) Easy access to intervention tool (physical opportunity), 2) Socially acceptable tool (social opportunity).	✓ Personalisation of intervention tool to encourage continued use and promote habitual use (automatic motivation).
✓ Automated instructions and guidance to ensure target behaviors are fully completed without error.	✓ Automated reminders to increase number of appropriate opportunities to complete target behavior (physical opportunity).	✓ Storage and presentation of information outlining why the intervention tool will be beneficial (reflective motivation).

Figure 6.1: COM-B Model utilized in designing features of the attentive eating app. From Robinson et al. (2013b).

Patrick et al. (2014) provides another good example of the use of theory to drive the design of a technical system. They utilized various behavioral theories in the design of the SMART (Social Mobile Approaches to Reduce WeighT) intervention. This included increasing physical activity as well as healthy eating. They utilized six different platforms: a social networking service, mobile apps, website with blogs, email, SMS, and occasional contact with a health coach as part of this multi-platform intervention. They drew heavily from previous literature and behavioral theories such as *Social Cognitive Theory* (Bandura,1986), *Control Theory* (Carver and Scheier, 1982), and *Social Network Theory* (Kadushin, 2004) to inform the design of their system, which focused primarily on theoretical constructs from the above mentioned theories, such as: self-monitoring, intention formation, goal-setting, feedback on performance, self-efficacy, social support, problem-solving, tailoring, and ecological support.

One example that incorporates a few of these theoretical constructs is their mobile app called "Goal Getter!"

Figure 6.2: An example of a theory-driven app (Goal Getter) used in the SMART intervention. From Patrick et al. (2014).

This app allows participants to self-monitor their behavior, set goals, and receive feedback related to progress toward achieving the goal (see Figure 6.2 for screenshot examples from their app). This is based on the premise that behavior change is mediated by improvements in self-efficacy, and self-efficacy, in turn, can be increased via strategies such as goal setting, self-monitoring, feedback on performance, and reviewing relevant goals. The goals are also encouraged to be specific, measurable, achievable, realistic, time-based, and enjoyable. Goals can also be shared via a social network service to encourage social support.

A Word of Caution about the Current State of Theories

A core reason why it is tricky to find the right theory is simply that many theories that are relevant for mobile user research are poorly specified, poorly evaluated and validated, and/or described to imply that they are true across a wide range of domains but are actually useful for a much more focused group of individuals in particular settings or for a particular problem. Based on this, there are a lot of "theories" and choosing the right one is very difficult. A recent expert-consensus study was conducted to create a "compendium" of theories about behavior and behavior change.[33] The group identified 83 unique theories, each with multiple constructs and proposed areas of interaction.

Beyond these points, there are also known shortcomings with current behavioral theories that any mobile user researcher attempting to use them should be aware of. These shortcomings often lead to common pitfalls that practitioners and behavioral scientists alike tend to fall into when utilizing behavioral theories (see Table 6.1 for highlights on these points that are described in more detail in Hekler et al. (2013a).

Taken in aggregate, it is important to acknowledge that it is a surprisingly difficult task to find the "right" theory for your domain. The following section delineates a strategy to attempt to find the "right" theory.

Table 6.1: Summary points of the common pitfalls and shortcomings of behavioral theories from Hekler et al. (2013a)

Section	Summary	Take-home Message
Common Pitfalls in Using Behavioral Theory	Common pitfalls include • ignoring context, • treating design guidelines as "fact," • falling prey to confirmation bias (i.e., believing what you expect to see), and • falling prey to type III error (i.e., concluding from null findings that a construct is not useful when, in fact, it was not designed well compared to the theory/construct and thus it was never actually tested).	It is easy and common for both behavioral scientists and HCI researchers to fall prey to the common pitfalls in using behavioral theory. Be ever wary of these when engaging with behavioral theories.

---

[33] See http://www.behaviourchangetheories.com/ {link verified Dec 28, 2016}.

| Shortcomings of Behavioral Theory | Current behavioral theories
• explain only a small portion of the variance,
• often cannot be falsified (and thus cannot be studied scientifically), and
• are fragmented with an over-abundance of behavioral theories, with limited utility. | While there are a few very good behavioral theories with solid empirical evidence (e.g., look into Applied Behavioral Analysis, which is the practitioner extension of behaviorist practices), most behavioral theories are best thought of as heuristics rather than as scientifically verified facts. This means that they should be used more like a logic-test compared to other HCI methods, not as a definitive "truth" that must be followed. |

6.4 SELECTING THE "RIGHT" THEORY(IES)

There are three general approaches we would like to highlight for selecting a theory. The first is to choose a theory that is familiar to you or your colleagues (e.g., the *Transtheoretical Model*, *Social Cognitive Theory*, and the *Health Belief Model*, are well-known theories). The second strategy is to utilize information about your target users (via previous scientific research about them or via your own user studies) to then find a theory that matches your research focus. The third strategy involves using meta-models to help guide some initial formative work and to then use this formative work to develop a list of contender conceptual frameworks to work from. We articulate how to use each approach and discuss advantages and disadvantages next.

6.4.1 USING A FAMILIAR THEORY

Often, the simplest strategy for picking a theory is to pick popular theories to work from. There are many advantages to this approach. Most apparent is that this is often the easiest way to engage with a theory as, implicitly, the thought is that popular theories are popular because they are "good" theories. This is an advantageous assumption as it often means that the mobile user researcher can pull from a large body of previous work when developing ideas from popular theories, particularly measures. In addition, using popular theories can be valuable for describing your work to others since colleagues may be aware of the theories and thus you can use shorthand for describing complicated concepts and inter-relationships (e.g., using the construct of self-efficacy rather than writing out confidence in one's ability to engage in a behavior). Finally, if you have previous experience with a theory, it often means that you have an understanding of the more subtle predictions that are made within the theory. As such, there is less of a "ramp up" for moving from just a rudimentary understanding into actual application of the theory.

While there are advantages to this approach, the primary disadvantage is that the theory you choose may not be well matched to the problem you are working on. This disadvantage is critical to be mindful of because it is very plausible to use a conceptual framework to guide all work, only to realize after much research that the conceptual framework has been obfuscating, and not illuminating, the truth about a given problem domain. Put more metaphorically, it is plausible for a behavioral theory to put you in the wrong "box" of inquiry, thus requiring the "outside the box" thinking so often colloquially desired. We emphasize that the choice of a particular theory should not be taken lightly or based purely on what is popular in general, but also because it fits your research problem well. A key strategy to use if you do end up starting with a popular theory is to develop strategies for testing assumptions that are implied by the theory to ensure you are not being led astray by the theory. For example, the TTM was originally developed for better understanding smoking cessation. While it has now been extended to a variety of other domains, there are many facets of the theory that implicitly hold over from this original focus on smoking cessation such as the length of time a person would likely be in each stage. A logical test of the assumption is to think through and actively challenge the more subtle predictions with your target user group to see if they hold up (e.g., will it really take a person two weeks, as predicted by TTM, to prepare to engage in being more physically active?).

6.4.2   UTILIZING USER INSIGHTS AND PREVIOUS RESEARCH

The second strategy for picking a theory is to utilize insights about your users. This can be done either via previous research, your own user studies, or both. Advantages include increased likelihood that you are building on a theory that will lead you in the right direction as it matches your research area. This is important to emphasize because it is often the case that theories seem appropriate at first, but after greater scrutiny, both with the user and via better understanding of the theory itself, the original chosen theory becomes increasingly less relevant. For example, imagine you were working on developing a system to support improved energy conservation. Through your user research you start to gain insights on the most impactful behaviors a person can engage in to improve their energy efficiency as your primary goal. In the back of your mind, however, you had heard of previous studies that used Cialdini's constructs of *descriptive norms* (i.e., perception on what others are actually doing) and *injunctive norms* (i.e., perception on what you perceive your social group believes is the right thing to do) to impact towel usage in a hotel to great effect (Goldstein et al., 2008). Upon further analysis with your group, however, you find that energy consumption is a highly idiosyncratic problem with many factors such as home size, local power company regulations, and overall home-efficiency factors, that allows for a very wide range of behaviors to occur to achieve the energy reduction. Further, you realized that impactful behaviors are those that happen very infrequently, such as buying a high-efficiency air conditioner or other appliance. This knowledge of your user and problem provides you with a logical foundation for recognizing that descriptive and injunctive norms may not be very useful in this context, but instead other factors focused on helping individuals understand the pros and cons of complex decisions (e.g., motivational interviewing perhaps) might be better for influencing impactful energy conservation behaviors (e.g., buy a high-efficiency refrigerator or air conditioner). It is quite plausible that starting with theory for a problem like this could lead the researcher astray if they were to continue to follow the theory without questioning its appropriateness for their particular problem.

A key disadvantage of this selection strategy, particularly when used alone and not as part of the hybrid third strategy discussed below, is that it often requires a great deal of previous knowledge about a wide range of behavioral theories. As the above example illustrates, a person experienced in behavioral theories can easily think through a variety of meta-models and conceptual frameworks to find the right match to the problem (e.g., moving from social norms to motivational interviewing). That is nearly impossible, though, for those with limited knowledge about theories, or even those with detailed knowledge of only a few theories. As such, for beginners in mobile user research, we do not suggest using this approach until you have a stronger background knowledge in behavioral theories. For mobile user researchers with knowledge of several different behavioral theories, starting explicitly with the user is a very good strategy as it allows for better explication of the problem, which can then guide theory selection to match the problem. For those with limited knowledge of behavioral theories, we suggest the third strategy, which is a hybrid of the first two.

6.4.3   META-MODEL FOLLOWED BY CONCEPTUAL FRAMEWORKS

The third strategy is to use a meta-model—which as a reminder, is a generalized framework for organizing thinking—to help define the initial formative work. Following this formative work, the goal is to then pick contender theories to work from. Indeed, an implicit goal of this approach is to be able to identify competing conceptual frameworks to use after completing the initial formative work. In our view, this third strategy of first using a meta-model for the initial formative work to identify contender conceptual frameworks is likely the most robust strategy for choosing theories. This is primarily because there are more "checks" within the process for ensuring the theories used are well matched to the problem being studied.

To illustrate our third strategy, let's imagine a researcher is working on developing a smartphone-based application to increase fruit and vegetable consumption in working adults. The first step is to identify an appropriate meta-model for understanding behavior. Based on prior knowledge of meta-models (note, we will go through several later in this chapter to establish this prior knowledge), the researcher chooses to use the COM-B model. The researcher then develops interview questions and observation strategies focused on better understanding working adults' capabilities, opportunities, and motivations to eat fruits and vegetables. Through this work, it becomes apparent that working adults have the capability to eat fruits and vegetables both at home and at work but often lack the motivation to do so, particularly when other more enticing food choices, such as fried foods or sweets, are available. Based on this, the researcher determines that *motivation* is the domain to target specifically and thus starts to identify theories about motivation. The researcher identifies theories such as *self-determination theory* (Ryan and Deci, 2000), *motivational interviewing* (Miller and Rollnick, 2012), and the *transtheoretical model* (Prochaska and DiClemente, 1994) as contender theories. The researcher then continues to do formative research, but now with better explicated constructs about motivation such as *intrinsic* vs. *extrinsic motivation* (i.e., having an internal drive to do a behavior vs. being rewarded externally for doing a behavior), *ambivalence* (i.e., the recognition that a person often has competing motivations such as the motivation to be healthy *but also* the motivation to enjoy candy), and the *stages of change* (from the TTM discussed earlier). The researcher can then examine which of these constructs, if any, seem most important for impacting their target users' fruit and vegetable consumption. The process then continues with this constant check between theoretical constructs and observational work with target users and eventually leads into concept sketching low- and high-fidelity prototypes[34] that represent the constructs (e.g., reminding a person of the positive health reasons to eat vegetables each morning via a motivational text message), and eventually a fully functional system.

Overall, the advantages of this approach are that it provides a strategy for focusing preliminary exploratory work while still also remaining broad enough to not "pigeonhole" into one particular way of understanding the problem too soon. In addition, it also still emphasizes the user as the core reference point for selecting a theory. Key disadvantages of this approach include that it likely requires more time as it implicitly emphasizes an iterative process.

6.5   JUDGING THE QUALITY OF A THEORY

Now that we have defined different possible ways to choose the "right" theory, we turn to a discussion on how best to judge the quality of any given theory you use. This can be very valuable to further gauge how much credence you, as the researcher, should place in any given theory. Note that in this instance, our goal is to provide you with strategies for evaluating the theory, in general, not evaluating if the theory is a good match for your particular problem. This is important and distinct from ensuring the theory is a good match for your target user group or problem, as it allows you to get a sense of how well the theory was developed.

A recent expert-consensus study was conducted to develop criteria for judging the quality of a theory (Michie et al., 2014, pp. 22–23). Results from this work are highlighted in Table 6.2. As Michie et al. succinctly stated, "*Good theories…begin with a parsimonious, coherent explanation of phenomena and generate predictions that can be compared against observation.*" While their nine criteria

---

[34]   Low- and high-fidelity prototypes are discussed in Chapter 3.

and questions (which are listed in Table 6.2) are valuable, we have slightly reformulated them into four questions that we believe are of particular importance for mobile user research. These are as follows:

1. Are the constructs and the relationships between constructs clearly specified?

2. Does the theory provide specific explanations for a given behavioral phenomenon that can be tested and falsified?

3. Does the theory specify its own boundary conditions (i.e., the times when the theory is NOT relevant)?

4. Is there previous work that you can draw from to find clear operational definitions (i.e., well specified instantiations, like specific code, that can be used as measures or as intervention strategies) of the constructs in your area of research?

The more times the answer is yes to these questions, the higher likelihood that the theory is of sufficient quality to be considered for use.

Table 6.2: ABC of behavior change theories (Michie et al., 2014)

The expert group agreed on nine criteria by which to assess the quality of a theory.

1. **Clarity of constructs:** "Has the case been made for the independence of constructs from each other?"

2. **Clarity of relationships between constructs:** "Are the relationships between constructs clearly specified?"

3. **Measurability:** "Is an explicit methodology for measuring the constructs given?"

4. **Testability:** "Has the theory been specified in such a way that it can be tested?"

5. **Being explanatory:** "Has the theory been used to explain/account for a set of observations?" (statistically or logically)

6. **Describing causality:** "Has the theory been used to describe mechanisms of change?"

7. **Achieving parsimony:** "Has the case for parsimony been made?"

8. **Generalizability:** "Have generalizations been investigated across behaviours, populations and contexts?"

9. **Having an evidence base:** "Is there empirical support for the propositions?"

6.6   A FEW THEORIES TO GET STARTED

Assuming you have little to no previous experience with theories, we now want to turn to some popular theories that can often play the role of both meta-models and conceptual frameworks. We highlight these not to suggest that they are the "right" theory for your problem, but more so to help illustrate how to think through the advantages and disadvantages of various theories.

As highlighted above when we defined the terms, the distinction between a meta-model and conceptual framework is largely artificial and based on the likelihood that the theory is providing a generalized structure that is true across domains vs. more specific to a smaller number of domains. To help provide a bit more structure, there are two other meta-models beyond the COM-B model that we want to highlight: Fogg's Behavioral Model and the Behavioral Ecological Model.

Fogg's Behavioral Model (FBM) asserts that for a target behavior to occur, a person needs to:

1. be sufficiently motivated;

2. possess the ability to perform the behavior; and

3. be triggered to perform the behavior.

These three factors *need to occur at the same moment*, else the behavior will not take place (Fogg, 2009). The FBM postulates that a person with high motivation *and* high ability is likely to perform the target behavior *when* it is triggered effectively at the right moment, and that for any behavior, a person must have a non-zero level of ability as well as motivation. Having said that, a person with high ability but low motivation may still perform the behavior when it is simple enough, and a person who ranks low on ability may still perform or at least attempt to perform the behavior if they are sufficiently motivated. The model emphasizes the element of "timing" of the trigger, or the "opportune moment to persuade." The opportune moment is considered to be any point above the "behavior activation threshold." The behavior activation threshold is reached when the appropriate combination of motivation and ability is achieved, and an appropriate trigger beyond this threshold will cause the target behavior to be carried out.

The COM-B meta-model and the FBM are just two of the meta-models that mobile user researchers could consider using in their work. Another meta-model that we suggest is the *Behavioral Ecological Model* (or "BEM") (Hovell et al., 2009). The BEM is a model that has a more contextual approach, in that it focuses on behavior as a function of the events in a person's environmental context, both current and past. It describes behavior as a result of two hierarchical interacting systems: factors within the skin (genetics, anatomy, physiology, learning history), and factors outside the skin (*individual level*—normative group, *local level*—clinical services, *community level*—laws, media, *social/cultural level*—nationality, culture specific). This model also considers behavior as being temporal in nature in the context of the above mentioned stimuli. Ecological models such as the BEM can provide a framework for comprehending these multiple levels of influence of the determinants of behavior, especially when the focus is on the population level.

There are some behavioral theories that have elements both of a meta-model as well as a conceptual framework. For example, *social cognitive theory* (or "SCT;" mentioned previously in the SMART project) includes, as part of its articulation, a high-level meta-model that is labeled "triadic determinism" (Bandura, 1986). Specifically, this part of SCT suggests that behavior, a person, and the environment influence one another (see Figure 6.3). This part of SCT functions as a meta-model as it is light on specifics but likely true across a wide range of behaviors and contexts. Other aspects of SCT, particularly the sub-part that is often labeled "self-efficacy theory," function more as a conceptual framework. We highlight this point as it is important to be mindful of the components of a theory that are general vs. specific.

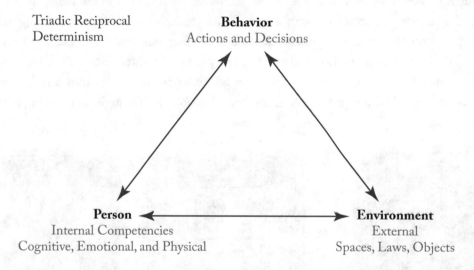

Figure 6.3: Triadic reciprocal determinism. Based on Bandura (1986).

Other theories that have both elements of a meta-model and conceptual framework include the *transtheoretical model* (Prochaska and DiClemente, 1994) (i.e., the stages of change is a bit more of a meta-model whereas the processes of change, which articulate very specific factors that occur within each stage to move a person from one stage to the next, are more functioning as a conceptual framework) and Lazarus and Folkman's *transactional model of stress and coping* (Lazarus and Folkman, 1987) (i.e., the general concept that stress, one's appraisal of stress and coping strategies for dealing with stress, and the external environment all co-interact to determine the impact of stress) could be a meta-model. We caution the overuse of these theories as meta-models, however, as they are more focused on domains. For example, the transtheoretical model is only useful when the target of the research is behavior change. For the transactional model of stress, it is most relevant when stressors are a key part of the problem area. Nonetheless, we wanted to highlight them here as contender meta-models. Please do remain mindful of the risk of specificity. The more specific the theory you choose, the increased likelihood that you may be led astray by it because it is not well matched to the particular problem that you are studying. Table 6.3 provides short descriptions of several theories that are relevant to mobile user research along with a few examples of their common use-cases.

6.7 AN ILLUSTRATIVE CASE STUDY: THE MILES STUDY

The Mobile Interventions for Lifestyle Exercise and eating at Stanford (MILES) study is an example of a study that utilized theory throughout the development and evaluation process (King et al., 2013, 2016). We developed three smartphone applications aimed at increasing physical activity and reducing sedentary behavior among older adults. The study involved tracking physical activity utilizing the built-in accelerometer within an Android smartphone and then translating those data about physical activity into three "motivational frames" that, building on Consolvo et al's UbiFit system, were displayed on the live wallpaper of the phone. The intervention strategies utilized in all three apps were based on previous literature on behavior-change theories and interventions, interviews conducted with potential users, and evidence obtained through users during their initial design process.

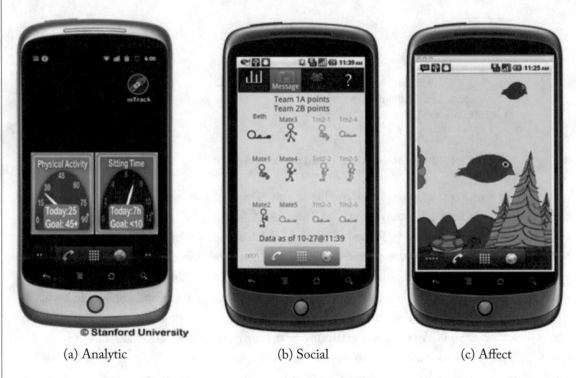

(a) Analytic	(b) Social	(c) Affect

Figure 6.4: Wallpaper graphics for the (a) Analytic, (b) Social, and (c) Affect apps. From King et al. (2013).

The first app, called the "Analytic" app, was based on *Social Cognitive Theory* (Bandura, 1986) and self-regulatory principles of behavior change (Umstattd et al., 2008). Based on these theories, the constructs that this app incorporated included:

- goal setting,

- behavioral feedback,

- problem-solving around barriers to change, and

- informational tips or advice.

In this app, participants were required to set weekly goals, aimed at increasing moderate-to-vigorous physical activity. The aim was to incorporate "graded" goals in order to increase self-efficacy (more on this below). Numeric feedback was provided to participants as the "live wallpaper" throughout the day (Figure 6.4a).

The second app, called the "Social" app, was based on social influence theory and perspectives (Cialdini and Goldstein, 2004). It aimed at incorporating theoretical constructs of:

- social support,

- social comparison,

- modeling of behaviors by similar individuals, and

- social normative feedback.

This was achieved through creating a virtual "avatar" for each participant, and assigning all participants to a particular virtual group. The feedback consisted of the avatars of other de-identified participants in the study displayed as the wallpaper of the phone (Figure 6.4b). The posture depicted by the avatar was symbolic of how active that participant had been up to that point each day. Features, such as a display of the participant's own physical activity for the day and the group average for the day, were included for social comparison.

The third app, called the "Affect" app, was based on operant conditioning principles (Skinner, 1953, 1981), particularly positive reinforcement. This app also incorporated the use of an avatar (in the form of a bird), but without the use of a social group. This avatar reflected the participant's current physical activity behavior, again, as a wallpaper displayed on the phone screen (Figure 6.4c). Positive reinforcement was achieved through incorporating "rewards" in the form of increasingly changing behaviors of the bird avatar (gives a thumbs up, appears happy, etc.) when the predetermined levels of physical activity were surpassed as well as "bird songs" that were sung while a person was walking and after meeting pre-specified thresholds that were based on a regimented reinforcement schedule (a key principle within operant conditioning, see Fisher et al. (2011)). Previously reported results (King et al., 2013) revealed that each app significantly increased physical activity. Further, as expected, different participants gravitated more to one type of app over the other; there was no "one size fits all" app that was "right" for everyone in the study.

With this overview as a backdrop, we now turn to a more detailed discussion on how theory informed the measures used and the design of the three apps. The research team was focused on promoting physical activity among mid-life and older adults. A review of the literature revealed that, at the time, very little research was currently available for promoting physical activity among this target group via mobile or "mHealth" devices such as smartphone apps, but there was some previous research that utilized personal digital assistants (PDAs) to promote physical activity that provided some insights on likely behavioral theories that could be valuable for this target group, particularly Social Cognitive Theory, Self-Determination Theory, and the Transtheoretical Model.

Next, the researchers conducted formative interviews to help better understand the specific needs of the target users. Through this work, the researchers learned that it appeared that Social Cognitive Theory, Self-Determination Theory, and the Transtheoretical Model did not appear to fully capture the various problems observed by the target group. In particular, it seemed that no one theory perfectly fit the reported problems and motivators mid-life and older adults might find most useful. Based on these observations, the research team engaged in iterative development, with the eventual design decision to build on different "motivational frames" that tap into the various needs of the target group. In particular, some individuals reported that they just wanted the numbers and help problem-solving. Others felt that they would only be motivated if they were being active with others. And still others felt that the act of "physical activity" was too strict and driven; these individuals desired activity to be made more "fun."

With motivational frames identified via interviews and prototyping, the research team then sought out theories that could both inform each "motivational frame" as well as build systems with appropriate theoretical fidelity (i.e., that the system design is in line with the predictions and suggestions made by a given theory or construct).

In this example, theory was used, in tandem with evaluating previous work and continued user-centered design work, to define what to measure and how to design the system. According to SCT, a strategy for increasing physical activity for this group, particularly those who were interested in "just the numbers," would be to help the users increase their self-efficacy in being active (i.e., confidence in their ability to be physically active). SCT suggests that a key strategy for increasing self-efficacy is by having individuals achieve small successes with engaging in any behavior. As such, a key system feature to support self-efficacy was via a goal-setting component for physical activity that could promote graded improvements in activity. One key tenant of Goal-Setting Theory, however, emphasizes the importance of fostering a sense of control and "ownership" over the goals set (Locke and Latham, 2002). This is often accomplished by allowing individuals to set their own goals. Based on these two theories, the team established two design constraints for good goal-setting: (1) provide a gradual improvement to facilitate self-efficacy; and (2) facilitate autonomy and "ownership" of a goal by providing choice.

Within the MILES "analytic" app, which focused explicitly on goal-setting, the team went through several iterations that balanced these two design requirements. For example, in the initial development work for MILES, early versions of the goal-setting feature placed greater emphasis on providing information, such as the national guidelines for physical activity recommendations, then allowing choice, with the assumption that individuals would naturally choose a gradual improvement. Specifically, early designs framed goal-setting as such, "Last week you were active xx min/day. How many minutes per week would you like to shoot for this week?" During our design work, it was found that, particularly during earlier weeks, many individuals did not choose gradual goals but instead often chose something such as the national guidelines, which were mentioned earlier in the interaction with the app. As such, the system was not facilitating graded goals, which is essential based on SCT Inspired by a behavioral economics concept of "choice architecture" and coupled with lessons from our initial user testing, the group ultimately decided upon a forced choice option, as it offered the desired gradual improvements while still facilitating autonomy. Specifically, the system tracked a person's previous week of behavior and then provided four options for the individual to choose from for the following weeks' activity:

1. +30 min/week from last week (note, this value was calculated for them, such as if a person was active 65 min, this option would be 95 min/week);

2. +60 min/week from last week;

3. +90 min/week; or

4. set a custom goal.

As this example implies, theory can both help to define what features to put in (i.e., graded goal-setting supported by good choice architecture), as well as the underlying purpose for the feature (i.e., to foster increased self-efficacy and autonomy). As can be seen through the illustrative example above, this can be very valuable when user testing occurs, as it allows the researchers to link the feature (i.e., graded goal-setting) to a specific measure (i.e., self-efficacy) to see if their feature is impacting the desired outcome (Klasjna et al., 2011). Further, theory, particularly an awareness of different theoretical constructs, can be very valuable for making new design decisions when desired outcomes are not occurring (e.g., the use of choice architecture to facilitate graded goal-setting).

6.8   SUMMARY

The goal of this chapter was to establish how mobile user research can use theory, provide pragmatic advice on how to engage with theory, particularly with limited prior knowledge, and to then provide illustrative examples of key points. While engaging with previously established theory can be daunting, we hope that we have provided actionable strategies for actively and effectively using theory. As we hope we've delineated throughout this chapter, there are great advantages to building on previous work that has been encapsulated in theory. As such, while it may be daunting at first, the rewards often come in the form of more useful and usable mobile systems.

Table 6.3: Examples of several theories that are relevant to mobile user research					
Theory	Description	Reference	Type of Theory*	Examples of Application / Use	Examples of common use-case
Goal-setting theory	This theory states that goal-setting affects behavior and discusses various mechanisms through which this occurs.	Locke and Latham (2002)	CF	Loock et al. (2013); Consolvo et al. (2009)	Organizational behavior research, health behavior (e.g., physical activity)
Self-efficacy theory	Self-efficacy is a person's confidence in their ability to perform a behavior. This theory discusses how self-efficacy affects behavior.	Bandura (1977)	CF	Chao et al. (2013)	Health behavior (smoking, alcohol use, physical activity, eating behavior etc.)
Self-determination theory (SDT)	The SDT is a theory of motivation. It theorizes about the various constructs that promote or impede the process of intrinsic motivation and other psychological processes.	Ryan and Deci (2000)	CF	Gustafson et al. (2014)	Health behavior (smoking cessation, medication adherence, chronic disease management, etc.), education parenting
Unified Theory of Acceptance and Use of Technology (UTAUT)	The UTAUT is a theory about the acceptance of information technology. It focuses on the determinants of intentions and usage of technologies and the moderators of those relationships.	Venkatesh et al. (2003)	CF	Ifenthaler and Schweinbenz (2013)	Information systems, information technology
Diffusion of Innovations	This theory describes the process whereby an innovation/new idea gets transmitted over time among the members of a social system.	Rogers (1962)	CF	Greenhalgh et al. (2008)	Sociology public health marketing communication studied

Theory	Description	Reference	Type of Theory*	Examples of Application / Use	Examples of common use-case
Transtheoretical Model (TTM)	The TTM is a stage-based model of behavior change that theorizes that individuals move through six stages of behavior change that each involve distinct processes that can be utilized to move from one stage to another.	Prochaska and Velicer (1997)	CF	Consolvo et al. (2009)	Health behavior research (physical activity, smoking cessation, etc.)
Prospect Theory	Prospect Theory is a behavioral economic theory about decision making and choosing between alternatives, especially those involving risk.	Kahneman and Tversky (1979)	CF	Spence and Pidgeon (2010)	Economics/ finance marketing
Social Cognitive Theory (SCT)	SCT hypothesizes that the behavior, the individual, and the environment reciprocally influence one another (triadic reciprocal determinism).	Bandura (1986)	MM	Patrick et al. (2014)	Individual/ community based health behavior research (smoking cessation, alcohol use, physical activity, chronic disease management and prevention, etc.)
COM-B Model	This meta-model posits that for any behavior to occur, there must be the appropriate balance of an individual's capability, opportunity, and motivation to engage in the behavior.	Michie et al. (2011)	MM	Jackson et al. (2014)	Health behavior research

Theory	Description	Reference	Type of Theory*	Examples of Application / Use	Examples of common use-case
Fogg's Behavioral Model	This meta model states that for a target behavior to occur, an individual needs to be sufficiently motivated, possess the ability to perform the behavior, and be triggered at an opportune moment to perform the behavior.	Fogg (2009)	MM	Rabbi et al. (2015)	Persuasive technology
*CF = Conceptual Framework; MM = Meta model					

CHAPTER 7

Big Challenges and Open Questions

Over the past two decades, as mobile devices have matured into their current state, researchers have explored many ways to understand how new systems fit into people's lives. Throughout this book, we have provided practical overviews of a wide variety of these methods, including app instrumentation; diverse contextual sensing; lab studies; field studies, including diary studies and experience sampling; experimental designs to test for change, and how theory can help. However, many open challenges remain in capturing user behavior, determining if a solution is producing the intended outcomes, and in trusting that a particular sample of study participants or data are representative of a broader population for which you might be aiming to create solutions.

This chapter will cover several open challenges and topics in mobile user research. We explore ways to incorporate mixed methods and surveys to triangulate findings from smaller scale, longitudinal studies, and propose future work to evaluate the innovations that mobile technology has brought to user study methods. We also explore how log data can be captured from broad audiences to quickly understand a domain, and how it can be combined with data from qualitative studies to explain the behavior that was observed in the logs. Furthermore, we explore new frontiers on the theory side, including changing experimental design to focus on the individual instead of averages over a larger population, as well as building adaptive theories that work for specific populations.

At the end of this chapter, you will be aware of some of the latest research trends on these topics and prepared to understand mobile user behavior in more nuanced ways than the earlier single methods have explored.

7.1　DIARY STUDIES AND EXPERIENCE SAMPLING

In Chapter 4, we discussed how to use Diary Studies and the Experience Sampling Method as a way to understand users, their environment, and their use of a technology in the wild, at times where direct participant observation is not required or appropriate. These methods excel at providing insight into how systems fit into the rhythms of daily live, and they provide details about how a person's context impacts their use of a system. As with all methods, these have several strengths and limitations. In this section, we describe some of the ways we, and other, researchers are working to mitigate limitations, and we propose future work to evaluate how innovations in how to administer these techniques may have affected traditional strengths and limitations.

7.1.1   TRIANGULATING DATA

Diary studies are almost always fairly small scale. It takes effort to get each participant onboarded to the study and aware of the mechanisms to provide feedback for the diary. Often, it also takes a great deal of effort to analyze the qualitative data from the diaries and make sense of the themes. Our studies have typically involved 10–12 participants (Bentley and Metcalf, 2007; Bentley et al., 2011) and at most 60 (Bentley et al., 2013b) in a very large collaborative effort with several institutions. While we can get deep qualitative data about use from these samples to help us understand how the system was used (or not used) during the study period, these types of methods tell us little about how a broader population might use such a system. Knowing this is extremely important if one is deciding to start a new business or fully develop a commercial system based on an early prototype. Regardless of how well you recruit, a sample of 10–12 is unlikely to truly represent your target population.

To mitigate these issues, we have been combining diary studies with larger, quantitative methods such as usage logs or surveys. The diaries get us the deep qualitative data that explain why people are behaving in a certain way with the system, while the logs or surveys can inform us of how representative each behavior is within a broader population. The logs or surveys enable us to say that a particular behavior occurs with a particular frequency, and the diary and interview data help us investigate people's intentions and perceptions.

One area where we have been using this method frequently is in understanding email use. Over the past year, we have fielded dozens of surveys and performed thousands of investigations into usage logs to explore what people were doing in their email programs. For example, we saw a high volume of deal/coupon messages being opened. We also ran a survey that found that 45% of our fairly U.S.-representative sample had actually used a coupon from email in the past week (Bentley et al., 2017a), and that a third of the coupon redemptions happened entirely on a mobile device. We then ran a diary study to explore this behavior in more detail, investigating the types of coupons that were used, how they were saved until they were used, and issues in finding and redeeming them. This combination of quantitative and qualitative data helped us to not only deeply understand specific use, but also helped us to see how representative particular qualitative observations were in the broader population.

When conducting a study like this, sometimes the qualitative data can prompt the need for more quantitative data. Perhaps you find an unexpected behavior, but want to know how prevalent it is in the broader population. While other times it can go the other way and you can find something from usage logs or a large-scale survey and want to dig deeper to better understand what's going on.

At Yahoo, we have found rapid survey platforms to be quite reliable for these types of questions. Findings from quick surveys run on platforms such as SurveyMonkey or Amazon Mechanical Turk have been within 10% of more expensive market research panels (Bentley et al., 2017a) for a wide range of behavioral questions that we have asked. Typically we will run surveys with 150-200 participants using a U.S. panel for quick feedback on if a particular behavior is rare, fairly common, or something that almost everyone does. Of course, you will need to determine if panels such as SurveyMonkey's or Amazon Mechanical Turk's is appropriate for your particular needs.

7.1.2   EVALUATING NEW EXPERIENCE SAMPLING TECHNIQUES

As we discussed in Chapter 4, the techniques for administering experience sampling questionnaires have matured and expanded with technology. We, and other researchers, have used mobile technology to develop experience sampling tools, for example, Barrett and Barrett's *ESP* (2001), Froehlich et al.'s *MyExperience* (2007), Carter et al.'s *Momento* (2007), Fetter and Gross' *PRIMIExperience* (2011), and *PACO* (the Personal Analytics COmpanion) (Evans, 2016). In a related line of work, Hsieh et al. (2008) explored how providing helpful visualizations to participants, in an effort to make responding to questionnaires more personally useful to them, might affect—and ideally improve—response rates.

The innovations to these techniques seem like they would help to further enhance the strengths of experience sampling and mitigate its limitations. For example, it makes sense that a system-triggered event-based alert *should* lead to less bias than relying on a participant to remember to trigger a questionnaire when they detect that an event of interest has occurred. However, we have found less research that has systematically and rigorously explored how these innovations actually affect strengths and limitations of experience sampling. For example, is there really the reduction in observer bias that experience sampling is known for if a participant realizes that the research team is watching their responses as they come in? This might occur, for example, if the researcher reaches out to the participant based on the participant's response to a questionnaire shortly after the participant completed the questionnaire. The research community would strongly benefit from rigorous evaluations of experience sampling's strengths and limitations when administered with recent innovations. Lathia et al.'s work (2013) is a good start.

7.1.3   FROM SENSORS TO USABLE INFORMATION

As explored in Chapter 2, mobile devices—phones, in particular—allow us to capture a wide variety of data about a person's context, activity, or device usage. Mobile sensing and instrumentation has given us a deeper view into people's interactions with technology and with each other at a scale that we have never been able to capture before. This provides exciting opportunities for understanding human behavior, but it also raises many important open questions.

In this section, we will explore some of these open questions including new means of opportunistic sensing, ways to understand logged data using qualitative data, and new ways to gather log data, especially if you do not control the system that you are trying to study.

Designing Silence

Mobile devices and the broader IoT increasingly use notifications (e.g., visual notes, sounds, vibrations, etc.) to encourage a user to respond to something (Voit et al., 2016; Weber et al., 2016; Mehrotra et al., 2015; Pielot et al., 2014; Tanner 2013; Cvach, 2012). For example, the experience sampling strategies highlighted in Chapter 4 rely heavily on notifications. Recent estimates indicate that people receive 65–100 notifications per day from smartphones (Mehrotra et al., 2015; Pielot et al., 2014). Not surprisingly, people have a tendency to ignore notifications over time, which can create serious problems for mobile user research (e.g., potentially missing alerts to take surveys in an experience sampling study or nudges from an app designed to elicit behavior change). Conceptually, this sets up the need for understanding the optimal times and places when a notification should be sent. Put more simply, there is a strong need to "design silence" into our mobile and related technologies. There is both conceptual and methodological work taking place to enable this design of silence but the work is still largely in its infancy as the sensors and related data (see Chapter 2) are only now reaching a level of sophistication that enables truly robust notification management.

There are good roots to this work within the HCI literature, particularly in the context of inferring moments, often in a work context, when a person might be willing to be interrupted, often called *interruptibility* (Fogarty et al., 2005). While a good starting point, mobile user research establishes new challenges because of the far more variable contexts with which notifications would be sent but also because notifications in a mobile context are often triggering a wider range of behavioral responses beyond just clicking or responding to something digitally. Within the confines of just choosing the right notification for a fully digital interaction (e.g., responding to a text message or email), some of the more recent data-driven prediction algorithms explored show real promise for notification decision-making (Mehrotra et al. 2015). In particular, results from this work indicate the importance of personalized over generic models, better predicted notification acceptance via the algorithms over user-defined rules, and the feasibility of using the algorithms via online learning, thus demonstrating the feasibility for real-time notification management via these methods. This, and more recent work, which attempt to make these rules interpretable to the user (Mehrotra et al., 2016), highlight the need for understanding the preferences and desires of people, in given moments, for different types of information to be delivered via notification or not. With that said, this work is still at its relative infancy and more work on fostering better notification management is in order (Lee et al., 2015).

For example, as discussed elsewhere (see Consolvo et al., 2012), mobile devices are increasingly used to foster non-digital behavioral responses (e.g., inspire a person to go for a walk, eat better, interact with other friends nearby) that occur in the real-world. This added complexity of attempting to inspire a non-digital behavioral response creates further complexities for choosing the right times and places for sending notifications. Conceptually, Nahum-Shani et al. (2015) discuss the concept of "just-in-time" adaptive interventions, which are interventions specifically designed to support behavior change by providing support adaptively only in the "just-in-time" states when and where a suggestion would be appreciated and acted upon. Specifically, they define just-in-time states as a state when a person would be receptive to an intervention (e.g., a notification suggesting a person goes for a walk) and have the opportunity to act on the suggested intervention idea (e.g., being able to walk when the message is received). We have expanded upon these specifications to define how one might be able to computationally model "just-in-time" states via machine learning, dynamical systems modeling, and the like via state-space formulations (Hekler et al., 2016a). Overall, we see a very fruitful line of work focused on designing silence into our mobile systems to support better notification management and also fostering non-digital behavioral responses. Of course, it's likely that notifications meant to encourage a user to take an action that's not part of a study vs. notifications used for alerting a participant in an experience sampling study may need to work differently.

Hyper Personalization

Personalizing the mobile experience is starting to develop into a mature part of the mobile context. An interesting emerging area of research is on devising strategies for further supporting this personalized experience. In many ways, this work starts with increased emphasis not only on attempting to find generally true insights and patterns within data to improve systems and experiences but to also use research methods and practices that take advantage of a person's own data to help themselves. Indeed, Deborah Estrin has coined the term "small data" to explicitly separate efforts of using a person's data for themselves from the current emphasis on "big data," which is more in line with the former type of work of finding general principles within data that can be applied more broadly (Estrin, 2014). There are many possible extensions and uses of this. For example, in MyBehavior (Rabbi et al., 2015), a multi-armed bandit approach was used to generate personalized behavioral recommendations related to physical activity and diet, with results suggesting that the system can produce meaningful change in both of these outcomes. An interesting area of future work is on further advancing this type of personalization, which we argue, will be supported not only with the data but also with updates to research methods and theories (see the next sections).

Combining Log Data with Qualitative Data

As we discussed in Chapter 2, log data can tell you what a user is doing, but often gives very little insight as to why. As "big data" and the analysis of taps and usage to determine design changes through A/B testing continue to dominate product decisions in many companies, it is important to step back and qualitatively explore why particular behaviors are occurring. Maybe a particular design change leads to increased usage because users are confused or are trying to turn off whatever was just turned on. Perhaps the reason that a feature is not used is because users literally never see it, not that it lacks utility or the ability to drive growth if better positioned.

Similar to the *Triangulating Data* section above, using multiple, complementary techniques can help. One method we have used to augment log analysis is in-lab eye-tracking. While traditionally a technique for desktop studies, mobile eye-tracking setups are now available where the participant wears a special set of glasses while using a mobile phone or tablet. These solutions have the added benefit of recording what the user is looking at while distracted from the device and have a large potential to be used in more contextual settings. By having users interact with your system while wearing these glasses, you can see which features catch their attention, and which controls or buttons never receive a gaze. This can greatly help when redesigning complex applications or nudging users to engage with particular features.

Often a quick interview or diary study can be enough to explain certain log data. After seeing unexpected results in log data, we often conduct a few quick interviews (sometimes even remotely over video) to see how people are using a particular feature. Through this, we've sometimes discovered errors in the instrumentation itself, where a particular path was not captured, or we've discovered alternate ways that participants have for accomplishing a particular goal. Moving beyond relying solely on log data to understand use or to prioritize features or specific buckets in an A/B test is critically important.

As mentioned above, the more sources of data you can find that point in the same direction, the more confident you can be in your conclusions. Log data, while often seen as a "ground truth," contains the assumptions of the developer or product manager who defined the instrumentation specification. Participants in a study that's focused on collecting qualitative data (e.g., a diary study) might be biased to use an application more than actual users would under non-study conditions. But together, qualitative data can serve to explain unanswered questions from logs, or even indicate where the logs are not collecting the full information needed to understand use. Logs also have the power to show where your one "odd" participant in a qualitative study might be representative of millions of actual users in terms of interaction behavior, and thus their experiences might need to be weighted more strongly. King and Churchill (2017) explore this topic more deeply, particularly the business and design decisions that can be made with triangulated data and combining A/B testing with more traditional user research.

New Ways to Gather Log Data

One of the common criticisms of studies that use captured log data from proprietary systems is that it is often the case that only the researchers affiliated with those companies can run these types of studies.

Over the past few years, we have been exploring ways to collect data about the use of large systems without having any special access to the data stored on proprietary servers. These studies have typically involved large-scale surveys where we collect usage data in the form of device logs, browser histories, or screenshots of past behaviors. Collecting a representative sample of these logs can allow us to understand the use of other systems.

In one study at Yahoo, we were interested in the amount and types of contacts in a typical user's mobile phone book, and their calling/messaging frequencies to different types of contacts (Bentley and Chen, 2015). We developed a small Android app that would allow participants to tag contacts with relationship types and then upload those contacts, with hashed names and contact details to our server, along with the times and durations/lengths of calls and messages with that contact. Within a day, we were able to collect logs from 200 diverse participants via Amazon's Mechanical Turk, containing over 65,000 contacts. This allowed us to see, for example, that the average user had more contacts who were not recognizable by name than contacts who they would consider to be "friends," and that over 80% of calls and text messages went to an average of just 5 people. This helped us to design new contact applications for Android as well as contact views in Yahoo Mail.

In another study at Yahoo, we were interested in link sharing behaviors over mobile messaging. However, since we were not part of a large mobile carrier, we did not have access to a large database of SMS messages, nor was our mobile Messenger in wide use at the time. Instead, we ran an online survey that asked participants to search their most commonly used messaging app for the string "http" and take screenshots of the resulting conversations (Bentley et al., 2016). Within an hour or two, we were able to collect 300 examples of link sharing from diverse users from across the U.S. We were able to see what types of links were shared (e.g., videos, articles, restaurants/venues, etc.) as well as collect relationship information about the sender/recipient through the survey instrument. With this data, we could quickly create new types of visualizations for previewing different types of links as well as better integrate search and messaging as a product (Bentley and Peesapati, 2017).

Other researchers have used Android Accessibility APIs to study the notifications that people receive (Shirzai et al., 2014) or the time of particular app openings and durations of use (Bohmer et al., 2011). Collecting usage data is possible, even if a researcher does not directly control the platform they wish to observe. While some analysis can only be performed with full datasets, data collection via mid-scale surveys can get researchers quite far in understanding user behavior in applications or other systems for which usage data is not directly available to them.

As mentioned elsewhere in this book, always check with the experts at your institution to ensure that you are abiding by the terms of service and other important rules, regulations, ethical considerations, and so on.

7.1.4   FROM "ON AVERAGE" TO USABLE EVIDENCE

As highlighted already in Chapter 5, there are myriad experimental designs that are available for mobile user researchers. This diversity creates both opportunities and challenges. On the opportunity side, the myriad designs enable a much wider range of options for supporting increasingly more robust research processes to occur within mobile user research. This is exciting as researchers do not necessarily need to be shackled to experimental designs of the past to do their work, which often would require considerably more resources compared to what the researcher would gain from the study. For example, running a fully powered randomized controlled trial is highly resource intensive and, often, creating a robust causal inference is secondary to the mobile user researchers' interest in developing novel, useful, and usable systems (Kay et al., 2016). Micro-randomized trials and other within-person or "single case" designs provide a strategy for fostering robust evaluation, but within the resources more commonly available to mobile user researchers (Klasnja et al., 2015, 2017). As such, the cost-benefit of research projects can be taken into account when devising an evaluation strategy. Further, there are real opportunities for gaining a far more subtle understanding of when, where, how, and why systems and the modules of systems work together for producing effects (Hekler et al., 2016a). All of this enables the possibility of "usable evidence" (Hekler et al., 2016a; Klasnja et al., 2017).

Within HCI research, and by extension mobile user research, there is a mismatch between the robust and subtle details one can learn from robust human-centered design, which can translate into novel systems that are well-matched to target user groups, and the experimental design that is used to evaluate these designs. Indeed, many of the subtle details that a designer carefully takes into account within rigorous mobile user research is systematically "balanced out" using between-person randomized controlled trials. An acknowledgement of a much wider range of experimental designs, particularly with the use of methods like the micro-randomized trial, offer real opportunities for better alignment between design and evaluation. In particular, these more advanced methods enable a highly resource efficient process whereby several questions important to a mobile user researcher can be answered simultaneously, including: (1) how well a system worked overall; (2) how well individual components of a system worked; (3) how individual components of a system interact,; (4) how to choose among alternative designs of a single component; and (5) how system use and effectiveness is affected by context and user characteristics. Answers to these questions thus enables far more "usable evidence," meaning evidence that is designed specifically to support and facilitate the decision-making process of the individual/group using the evidence. Usable evidence can be contrasted with more "on average" evidence generated by trials such as between-person randomized controlled trials. While these trials do give insights on if a system works in general, a great deal of the subtly required to support decision-making for specific individuals and users is systematically lost and thus is not nearly as usable. We see great opportunities for a revolution in HCI and mobile user research with this movement toward methods that enable usable evidence (Hekler et al., 2016a; Klasnja et al., 2017).

7.1.5 EMPOWERING END-USERS IN PERSONALIZATION OF MOBILE EXPERIENCES

The promise of personalization also creates challenges for supporting said personalization; end-user programming of personalized mobile experiences is, thus, an interesting and plausibly future area of research that would benefit from these more advanced research methods. In particular, there is increased interest in personal informatics/quantified self as a sub-focus within mobile user research (Choe et al., 2014). As a further extension of this, there has been a growing interest in supporting self-experimentation of individuals that takes advantage of mobile/quantified self data to support individuals in personalized decision-making (Kravitz et al., 2014; Karkar et al., 2015; Daskalova et al., 2016; Lee et al., 2017). We see great opportunities in further supporting and advancing this area of work.

7.1.6 FROM THEORIES TO COMPUTATIONAL MODELS

As delineated in Chapter 6, there are myriad theories currently available to support mobile user research, but similar to the problem with experimental designs, the theories are often so abstract and general that they are not particularly usable for supporting decision-making within mobile contexts. This is problematic because, as discussed in Chapter 1 and throughout, mobile technologies are meant to be used whenever and wherever we go. As such, they need to be designed and studied in a way that takes their many contexts and uses into account. Advances in research methods, including experimental designs and data analysis strategies, and also in sensors (Chapter 2) are aligning to enable a major shift in how we create, evaluate, and iterate upon the "theories" that one can use for guiding mobile user research (Hekler et al., 2013a, 2013b, 2016b; Spruijt-Metz et al., 2015; Nahum-Shani et al., 2015; Patrick et al., 2016). It has long been known that human behavior (and, by extension, a clear understanding of how users will interact with mobile systems) is a complex phenomenon to study, with a variety of factors interacting with one another to inspire different behavioral responses (Hekler et al., 2016a). While this complexity has long been known, the theories that were created and used to help understand human behavior have largely been relatively static, simplified representations of this complex subject matter (Riley et al., 2011). There is a strong need for "theories" that match the complexity of human behavior, which has important implications for mobile user research.

A plausible roadmap for the future of understanding human behavior, which is made possible through mobile user research, is to look toward meteorology as what mobile user research could become (Patrick et al., 2016). It is important to realize that chaos theory grew out of the study of the weather, thus acknowledging the immense complexity involved in predicting the weather. Even with that level of complexity, it is an often overlooked and under-appreciated miracle of modern science that one can now simply take out their phone and learn about the future (in terms of weather) for their area. While the predictions are not perfect, they are most definitely at a level of precision that inspires countless people across the globe to use those predictions in making decisions daily (e.g., what should I wear?). Interestingly, many of the technological requirements that enable meteorology (e.g., better data that was standardized, connectivity of data across different types of data types, improved computational power) are increasingly coming into place for mobile user research. As such, there are very real possibilities for moving our theories from the static and largely descriptive models that are currently used to dynamic and predictive models that, like our experimental designs, can be used to support improved decision-making. Overall, this can enable the type of personalization and support delineated earlier in this chapter. As such, we advocate for a movement toward increased use of computational models over merely descriptive theories, whenever possible, within mobile user research, to continue to move our field toward more personalized, useful, usable, and enjoyable mobile systems (Hekler et al., 2016b).

7.2   SUMMARY

This chapter covered several open challenges and topics in mobile user research, including ways to mitigate methodological weaknesses by using multiple, complementary methods, and opportunities for future work including evaluating the innovations that mobile technology has brought to user study methods, changing experimental design to focus on the individual instead of averages over a larger population, and building adaptive theories that work for specific populations.

A decade into the smartphone era, we are still uncovering better ways to measure and understand a user's experience with a particular application or service. Now that you have read this chapter, we hope that you are more aware of some of the questions that we, as a research community, have yet to answer.

References

Abowd, G. D. and Mynatt, E. D. (2000). "Charting past, present, and future research in ubiquitous computing." *ACM Transactions on Computer-Human Interaction*, 7(1), 29-58. DOI: 10.1145/344949.344988. 2

Adalı, S. and Golbeck, J. (2014). "Predicting personality with social behavior: a comparative study." *Social Network Analysis and Mining*, 4(1), 1-20. DOI: 10.1007/s13278-014-0159-7. 20

Adams, M. A., Sallis, J. F., Norman, G. J., Hovell, M. F., Hekler, E. B., and Perata, E. (2013). "An adaptive physical activity intervention for overweight adults: a randomized controlled trial." *PloS one*, 8(12), e82901. DOI: 10.1371/journal.pone.0082901. 119

Ajzen, I. (1991). "The theory of planned behavior." *Organizational Behavior and Human Decision Processes*, 50(2), 179-211. DOI: 10.1016/0749-5978(91)90020-T. 134

Albert, W. and Tullis, T. (2013). *Measuring the User Experience: Collecting, Analyzing, and Presenting Usability Metrics*. Newnes. 4

Almuhimedi, H., Felt, A.P., Reeder, R. W., and Consolvo, S. (2014). "Your reputation precedes you: Reputation and the chrome malware warning." *Proceedings of the Symposium on Usable Privacy and Security: SOUPS '14*. 121

Ames, M. and Naaman, M. (2007). "Why we tag: motivations for annotation in mobile and online media." In *Proceedings of the SIGCHI Conference on Human Factors in Computing Systems* (pp. 971-80). ACM. DOI: 10.1145/1240624.1240772. 36

Amft, O. and Tröster, G. (2009). "On-body sensing solutions for automatic dietary monitoring." *IEEE Pervasive Computing*, 8(2), 62-70. DOI: 10.1109/MPRV.2009.32. 19

Anderson, J. and Rainie, L. (2014). "The Internet of Things will thrive by 2025: Many experts say the rise of embedded and wearable computing will bring the next revolution in digital technology." *Pew Research Center Report: Internet and Tech*; available from http://www.pewinternet.org/2014/05/14/internet-of-things/ {link verified 27 Nov 2016}. 1

Ariely, D. (2008). *Predictably Irrational*. HarperCollins:New York. 10

Aviv, A. J., Gibson, K., Mossop, E., Blaze, M., and Smith, J. M. (2010). "Smudge attacks on smartphone touch screens." *Proceedings of the 4th USENIX Conference on Offensive Technologies*. 62

Bandura, A. (1977). "Self-efficacy: toward a unifying theory of behavioral change." *Psychological Review*, 84(2), 191. DOI: 10.1037/0033-295X.84.2.191. 134, 138, 140, 148, 154

Bandura, A. (1986). *Social Foundations of Thought and Action: A Social Cognitive Theory*. Prentice Hall, Englewood Cliffs, NJ. 149, 150, 155

Bao, L. and Intille, S. S. (2004). "Activity recognition from user-annotated acceleration data." In *International Conference on Pervasive Computing* (pp. 1-17). Springer Berlin Heidelberg. DOI: 10.1007/978-3-540-24646-6_1. 16

Barrett, L. F. and Barrett, D. J. (2001). "An introduction to computerized experience sampling in psychology." *Social Sciences Computer Review*, 19(2), 175-85. DOI: 10.1177/089443930101900204. 83, 84, 85, 159

Beckmann, C. and Consolvo, S. (2003). "Sensor configuration tool for end-users: Low-fidelity prototype evaluation #1." *Intel Research Seattle Tech Report IRS-TR-03-009*, (July 2003). 65, 66, 67

Beckmann, C., Consolvo, S., and LaMarca, A. (2004). "Some assembly required: Supporting end-user sensor installation in domestic ubiquitous computing environments," *Proceedings of the 6th Int'l Conference on Ubiquitous Computing: UbiComp '04*, pp. 107-24. DOI: 10.1007/978-3-540-30119-6_7. 64, 68, 69

Beebe, J. (2001). *Rapid Assessment Process: An Introduction*. AltaMira Press. 3, 43

Beebe, J. (2005). *Rapid Assessment Process. Encyclopedia of Social Measurement*. Elsevier, Vol. 3, pp. 285-91. DOI: 10.1016/B0-12-369398-5/00562-4. 43

Bentley, F., Metcalf, C., and Harboe, G. (2006). "Personal vs. commercial content: the similarities between consumer use of photos and music," *Proceedings of the SIGCHI Conference on Human Factors in Computing Systems: CHI '06*, pp. 667-76. DOI: 10.1145/1124772.1124871. 50

Bentley, F. and Metcalf C. (2007). "Sharing motion information with close family and friends." In *Proceedings of the SIGCHI Conference on Human Factors in Computing Systems* 2007. pp. 1361-70. DOI: 10.1145/1240624.1240831. 36, 158

Bentley, F. R., Basapur, S., and Chowdhury, S. K. (2011). "Promoting intergenerational communication through location-based asynchronous video communication." In *Proceedings of the 13th International Conference on Ubiquitous Computing* (pp. 31-40). ACM. DOI: 10.1145/2030112.2030117. 49, 54, 81, 158

Bentley, F. and Barrett, E. (2012). *Building Mobile Experiences*. MIT Press. 7, 76

Bentley, F. and Basapur, S. (2012). "StoryPlace.Me: the path from studying elder communication to a public location-based video service," *CHI '12 Extended Abstracts on Human Factors in Computing Systems*, pp. 777-92. 47, 48, 83

Bentley, F. R., Harboe, G. F., Metcalf, C. J., Romano, G. G., and Thakkar, V. V. (2013a). "Multimedia device for providing access to media content." U.S. Patent No. 8,560,553. 51

Bentley, F., Tollmar, K., Stephenson, P., Levy, L., Jones, B., Robertson, S., Price, E., Catrambone, R., and Wilson, J. (2013b). "Health mashups: Presenting statistical patterns between wellbeing data and context in natural language to promote behavior change." *ACM Transactions of Computor-Human Interaction*, 20(5), Article 30. DOI: 10.1145/2503823. 23, 25, 26, 29, 158

Bentley, F., Church, K., Harrison, B., Lyons, K., and Rafalow, M. (2015). "Three hours a day: Understanding current teen practices of smartphone application use." ArXiv. http://arxiv.org/pdf/1510.05192. 27, 80

Bentley, F. R. and Chen, Y-Y. (2015). "The composition and use of modern mobile phonebooks." In *Proceedings of the 33rd Annual ACM Conference on Human Factors in Computing Systems: CHI '15*. DOI: 10.1145/2702123.2702182. 163

Bentley, F., Peesapati, T., and Church, K. (2016) "I thought she would like to read it: Exploring sharing behaviors in the context of declining mobile web use." In *Proceedings of the SIGCHI Conference on Human Factors in Computing Systems 2016*. DOI: 10.1145/2858036.2858056. 27, 163

Bentley, F. R., Daskalova, N., and Andalibi, N. (2017a). "If a person is emailing you, it just doesn't make sense: Exploring changing consumer behaviors in email." In *Proceedings of the 2017 CHI Conference on Human Factors in Computing Systems:CHI '17*. 158

Bentley, F. R. and Peesapati, S. T. (2017). "SearchMessenger: Exploring the use of search and card sharing in a messaging application." In *Proceedings of CSCW 2017*. DOI: 10.1145/2998181.2998255. 163

Bentley, F. R., Daskalova, N., and White, B. (2017b). "Comparing the reliability of Amazon Mechanical Turk and Survey Monkey to traditional market research surveys." In *Extended Abstracts of CHI 2017*, Case Study. ACM.

Beyer, H. and Holtzblatt, K. (1998). *Contextual Design: Defining Customer-Centered Systems*. Morgan Kaufmann Publishers. 5, 7, 43, 46, 76

Biglan, A., Ary, D., and Wagenaar, A. C. (2000). "The value of interrupted time-series experiments for community intervention research." *Prevention Science*, 1, 31-49. DOI: 10.1023/A:1010024016308. 113

Birks, M. and Mills, J. (2011). *Grounded Theory: A Practical Guide*. Sage Publications. 45

Böhmer, M., Hecht, B., Schöning, J., Krüger, A., and Bauer, G. (2011) "Falling asleep with Angry Birds, Facebook and Kindle: a large scale study on mobile application usage." In *Proceedings of the 13th International Conference on Human Computer Interaction with Mobile Devices and Services*, pp. 47-56. ACM. DOI: 10.1145/2037373.2037383. 26, 163

Bot, B. M., Suver, C., Neto, E. C., Kellen, M., Klein, A., Bare, C., Doerr, M., Pratap, A., Wilbanks, J., Dorsey, E. R., Friend, S. H., and Trister, A. D. (2016). "The mPower study, Parkinson disease mobile data collected using ResearchKit." *Scientific Data*, 3. DOI: 10.1038/sdata.2016.11. 28

Brookshire, B. (2013). "Psychology is WEIRD," Slate, http://www.slate.com/articles/health_and_science/science/2013/05/weird_psychology_social_science_researchers_rely_too_much_on_western_college.html {link verified Dec 28, 2016}. 131

Brownstein, J. S., Freifeld, C. C., and Madoff, L. C. (2009). "Digital disease detection—harnessing the Web for public health surveillance." *New England Journal of Medicine*, 360(21), 2153-7. DOI: 10.1056/NEJMp0900702. 20

Brownstein, J. S., Green, T. C., Cassidy, T. A., and Butler, S. F. (2010). "Geographic information systems and pharmacoepidemiology: using spatial cluster detection to monitor local patterns of prescription opioid abuse." *Pharmacoepidemiology and Drug Safety*, 19(6), 627-37. DOI: 10.1002/pds.1939. 28

Burnard, Philip. (1991). "A method of analysing interview transcripts in qualitative research." *Nurse Education Today*, 11(6), 461-6. DOI: 10.1016/0260-6917(91)90009-Y. 45

Burns, M. N., Begale, M., Duffecy, J., Gergle, D., Karr, C. J., Giangrande, E., and Mohr, D. C. (2011). "Harnessing context sensing to develop a mobile intervention for depression." *Journal of Medical Internet Research*, 13(3). DOI: 10.2196/jmir.1838. 111

Callegaro, M. (2017). "Recent books and journals articles in public opinion, survey methods, survey statistics, and big data." 2016 Update. *Survey Practice*, 10(1). 6

Campbell, D. T. and Stanley, J. C. (1966). *Experimental and Quasi-Experimental Designs for Research*. Houghton Mifflin. 102

Canzian, L. and Musolesi, M. (2015). "Trajectories of depression: unobtrusive monitoring of depressive states by means of smartphone mobility traces analysis." In *Proceedings of the 2015 ACM International Joint Conference on Pervasive and Ubiquitous Computing*, pp. 1293-1304. ACM. DOI: 10.1145/2750858.2805845. 19

Card, S. K., English, W. K., and Burr, B. J. (1978). "Evaluation of mouse, rate-controlled isometric joystick, step keys, and text keys for text selection on a CRT." *Ergonomics* 21(8), 601–13. DOI: 10.1080/00140137808931762. 1

Carter, S., Mankoff, J., and Heer, J. (2007). "Momento: Support for situated ubicomp experimentation." *Proceedings of the Conference on Human Factors in Computing Systems: CHI '07*, pp. 125-34. DOI: 10.1145/1240624.1240644. 87, 159

Carver, C. S. and Scheier, M. F. (1982). "Control theory: A useful conceptual framework for personality–social, clinical, and health psychology. "*Psychological Bulletin*, 92(1), 111. DOI: 10.1037/0033-2909.92.1.111. 140

Chao, Y. Y., Scherer, Y. K., Wu, Y. W., Lucke, K. T., and Montgomery, C. A. (2013). "The feasibility of an intervention combining self-efficacy theory and Wii Fit exergames in assisted living residents: A pilot study." *Geriatric Nursing*, 34(5), 377-82. DOI: 10.1016/j.gerinurse.2013.05.006. 154

Choe, E. K., Consolvo, S., Jung, J., Harrison, B., Patel, S. N., and Kientz, J. A. (2012). "Investigating receptiveness to sensing and inference in the home using sensor proxics," *Proceedings of the ACM Conference on Ubiquitous Computing: UbiComp '12*, pp.61-70. DOI: 10.1145/2370216.2370226. 78

Choe, E. K., Lee, N. B., Lee, B., Pratt, W., and Kientz, J. A. (2014). "Understanding quantified-selfers' practices in collecting and exploring personal data." In *Proceedings of the 32nd Annual ACM Conference on Human Factors in Computing Systems* (pp. 1143-52). ACM. DOI: 10.1145/2556288.2557372. 165

Choudhury, T., Consolvo, S., Harrison, B., Hightower, J., LaMarca, A., LeGrand, L., Rahimi, A., Rea, A., Borriello, G., Hemingway, B., Klasnja, P., Koscher, K., Landay, J. A., Lester, J., Wyatt, D., and Haehnel, D. (2008). "The mobile sensing platform: An embedded activity recognition system." *IEEE Pervasive Computing*, 7(2), 32-41. DOI: 10.1109/MPRV.2008.39. 19, 23, 98

Cialdini, R. B. and Goldstein, N. J. (2004). "Social influence: Compliance and conformity." *Annual Review of Psychology*, 55, 591-621. DOI: 10.1146/annurev.psych.55.090902.142015. 151

Colantonio, S., Coppini, G., Germanese, D., Giorgi, D., Magrini, M., Marraccini, P., Martinelli, M., Morales, M. A., Pascali, M. A., Raccichini, G., Righi, M., and Salvetti, O. (2015). "A smart mirror to promote a healthy lifestyle." *Biosystems Engineering*, 138, 33-43. DOI: 10.1016/j.biosystemseng.2015.06.008. 21

Collins, L. M., Murphy, S. A., Nair, V. N., and Strecher, V. J. (2005). "A strategy for optimizing and evaluating behavioral interventions." *Annals of Behavioral Medicine*, 30(1), 65-73. DOI: 10.1207/s15324796abm3001_8. 123

Collins, L. M., Murphy, S. A., and Strecher, V. (2007). "The multiphase optimization strategy (MOST) and the sequential multiple assignment randomized trial (SMART): new methods for more potent eHealth interventions." *American Journal of Preventive Medicine*, 32(5), S112-8. DOI: 10.1016/j.amepre.2007.01.022. 122

Consolvo, S., Arnstein, L., and Franza, B.R. (2002). "User study techniques in the design and evaluation of a Ubicomp environment," *Proceedings of the 4th International Conference on Ubiquitous Computing: UbiComp '04*, Springer-Verlag Berlin Heidelberg, pp. 73-90. DOI: 10.1007/3-540-45809-3_6. 45

Consolvo, S. and Walker, M. (2003). "Using the experience sampling method to evaluate ubicomp applications," *IEEE Pervasive Computing Magazine*, 24-31. 84, 86, 87, 95, 97

Consolvo, S., Roessler, P., and Shelton, B. (2004a). "The CareNet display: Lessons learned from an in home evaluation of an ambient display," *Proceedings of the 6th International Conference on Ubiquitous Computing: UbiComp '06*, Springer-Verlag Berlin Heidelberg, pp. 1-17. DOI: 10.1007/978-3-540-30119-6_1. 57

Consolvo, S., Roessler, P., Shelton, B.E., LaMarca, A., Schilit, B., and Bly, S. (2004b). "Technology for care networks of elders," *IEEE Pervasive Computing Mobile and Ubiquitous Systems: Successful Aging*, 3(2), 22-9. 58

Consolvo, S. and Towle, J. (2005). "Evaluating an ambient display for the home," *CHI '05 Extended Abstracts on Human Factors in Computing Systems*, pp. 1304-7. DOI: 10.1145/1056808.1056902. 57, 59, 60, 93

Consolvo, S., Smith, I.E., Matthews, T., LaMarca, A., Tabert, J., and Powledge, P. (2005). "Location disclosure to social relations: Why, when, and what people want to share," *Proceedings of the Conference on Human Factors in Computing Systems: CHI '05*, pp. 81-90. DOI: 10.1145/1054972.1054985. 84, 87

Consolvo, S., Harrison, B., Smith, I., Chen, M. Y., Everitt, K., Froehlich, J., and Landay, J. A. (2007). "Conducting in-situ evaluations for and with ubiquitous computing technologies," *International Journal of Human-Computer Interaction*, 22(1-2), 103-18. DOI: 10.1080/10447310709336957. 2

Consolvo, S., McDonald, D. W., Toscos, T., Chen, M. Y., Froehlich, J., Harrison, B., Klasnja, P., LaMarca, A., LeGrand, L., Libby, R., Smith, I. U., and Landay, J. A. (2008a). "Activity sensing in the wild: a field trial of ubifit garden." In *Proceedings of the SIG-*

CHI Conference on Human Factors in Computing Systems, pp. 1797-1806. ACM. DOI: 10.1145/1357054.1357335. 28, 87, 98, 139, 155

Consolvo, S., Klasnja, P., McDonald, D. W., Avrahami, D., Froehlich, J., LeGrand, L., Libby, R., Mosher, K., and Landay, J. A. (2008b). "Flowers or a robot army: Encouraging awareness and activity with personal, mobile displays," *Proceedings of the International Conference on Ubiquitous Computing: UbiComp '08*, pp. 54-63. DOI: 10.1145/1409635.1409644. 28, 87, 98, 101, 121, 139, 155

Consolvo, S., McDonald, D. W., and Landay, J. A. (2009). "Theory-driven design strategies for technologies that support behavior change in everyday life." In *Proceedings of the SIG-CHI Conference on Human Factors in Computing Systems* (pp. 405-14). ACM. DOI: 10.1145/1518701.1518766. 134, 154

Consolvo, S., Klasnja, P., McDonald, D. W., and Landay, J. A. (2012). "Designing for healthy life-styles: Design considerations for mobile technologies to encourage consumer health and wellness." *Foundations and Trends in Human-Computer Interaction*, 6(3-4), 167-315. DOI: 10.1561/1100000040. 160

Consolvo, S., Klasnja, P., McDonald, D. W., and Landay, J. A. (2014). "Designing for healthy life-styles: Design considerations for mobile technologies to encourage consumer health and wellness," *Foundations and Trends in Human-Computer Interaction*, 6(3-4), 167-315. 53, 87, 98, 101, 121

Cook, T. D. and Campbell, D. T. (1979). *Quasi-experimentation: Design and Analysis for Field Settings*. Rand McNally. 102

Cronbach, L. J. (1957). "The two disciplines of scientific psychology." *American Psychologist*, 12(11), 671. DOI: 10.1037/h0043943. 102

Csikszentmihalyi, M. and Larson, R. (1987). "Validity and reliability of the experience-sampling method," *The Journal of Nervous and Mental Disease*, 175(9), 526-36. DOI: 10.1097/00005053-198709000-00004. 83, 84

Cushing, C. C., Jensen, C. D., and Steele, R. G. (2010). "An evaluation of a personal electronic device to enhance self-monitoring adherence in a pediatric weight management program using a multiple baseline design." *Journal of Pediatric Psychology*, jsq074. 113, 114

Cvach M. (2012). "Monitor alarm fatigue: An integrative review." *Biomedical Instrumentation and Technology*, 46(4), 268-77. DOI: 10.2345/0899-8205-46.4.268. 160

Dallery, J., Cassidy, R. N., and Raiff, B. R. (2013). "Single-case experimental designs to evaluate novel technology-based health interventions." *Journal of Medical Internet Research*, 15(2). DOI: 10.2196/jmir.2227. 112, 113

Dallery, J., Glenn, I. M., and Raiff, B. R. (2007). "An Internet-based abstinence reinforcement treatment for cigarette smoking." *Drug and Alcohol Dependence*, 86(2), 230-38. DOI: 10.1016/j.drugalcdep.2006.06.013. 112, 113

Darrell, T., Tollmar, K., Bentley, F., Checka, N., Morency, L. P., Rahimi, A., and Oh, A. (2002). "Face-responsive interfaces: from direct manipulation to perceptive presence," *Proceedings of the 4th International Conference on Ubiquitous Computing: UbiComp '02*, Springer, Berlin Heidelberg, pp. 135-51. DOI: 10.1007/3-540-45809-3_10. 56

Daskalova, N., Metaxa-Kakavouli, D., Tran, A., Nugent, N., Boergers, J., McGeary, J., and Huang, J. (2016). "SleepCoacher: A personalized automated self-experimentation system for sleep recommendations." In *Proceedings of the 29th Annual Symposium on User Interface Software and Technology* (pp. 347-58). ACM. DOI: 10.1145/2984511.2984534. 165

Davis, F. D. (1989). "Perceived usefulness, perceived ease of use, and user acceptance of information technology." *MIS Quarterly*, 319-40. DOI: 10.2307/249008. 134, 139

Denzin, N.K. and Lincoln, Y.S. (2011). *The SAGE Handbook of Qualitative Research*, 4th ed. SAGE Publications. 3

de Vries, M., Dijkman-Caes, C. and Delespaul, P. (1990). "The sampling of experience: A method of measuring co-occurrence of anxiety and depression in daily life." In J. D. Maser and C. R. Cloninger (Eds.), *Comorbidity of Mood and Anxiety Disorders* (pp. 707-26), American Psychiatric Press:Washington, DC. 84

Dey, A. K., Wac, K., Ferreira, D., Tassini, K., Hong, J-H., and Ramos, J. (2011). "Getting closer: an empirical investigation of the proximity of user to their smart phones." In *Proceedings of the 13th International Conference on Ubiquitous Computing*, pp. 163-172. ACM. DOI: 10.1145/2030112.2030135. 32

Dunton, G. F., Dzubur, E., and Intille, S. (2016). "Feasibility and performance test of a real-time sensor-informed context-sensitive ecological momentary assessment to capture physical activity." *Journal of Medical Internet Research*, 18(6), e106. DOI: 10.2196/jmir.5398. 28

Enev, M., Gupta, S., Kohno, T., and Patel, S.N. (2011). "Televisions, video privacy, and powerline electromagnetic interference," *Proceedings of the 18th ACM Conference on Computer and Communications Security: CCS '11*, pp. 537-50. DOI: 10.1145/2046707.2046770. 3

Ertin, E., Stohs, N., Kumar, S., Raij, A., al'Absi, M., and Shah, S. (2011). "AutoSense: unobtrusively wearable sensor suite for inferring the onset, causality, and consequences of stress in the field." In *Proceedings of the 9th ACM Conference on Embedded Networked Sensor Systems* (pp. 274-87). ACM. DOI: 10.1145/2070942.2070970. 31

Estrin, D. (2014). "Small data, where n=me." *Communications of the ACM*, 57(4), 32-4. DOI: 10.1145/2580944. 20, 23, 30, 161

Evans, B. (2016). "Paco-applying computational methods to scale qualitative methods," *Ethnographic Praxis in Industry Conference*. DOI: 10.1111/1559-8918.2016.01095. 87, 159

Fan, Y., Wolfson, J., Adomavicius, G., Vardhan Das, K., Khandelwal, Y., and Kang, J. (2015). "Smar-TrAC: A smartphone solution for context-aware travel and activity capturing." Center for Transportation Studies University of Minnesota. 19

Felt, A. P., Ainslie, A., Reeder, R. W., Consolvo, S., Thyagaraja, S., Bettes, A., Harris, H., and Grimes, J. (2015). "Improving SSL warnings: Comprehension and adherence," *Proceedings of the Conference on Human Factors in Computing Systems: CHI '15*. DOI: 10.1145/2702123.2702442. 121

Felt, A. P., Reeder, R. W., Ainslie, A., Harris, H., Walker, M., Thompson, C., Acer, M. E., Morant, E., and Consolvo, S. (2016). "Rethinking connection security indicators," *Proceedings of the Symposium on Usable Privacy and Security: SOUPS '16*. 121

Felt, A. P., Reeder, R. W., Almuhimedi, H., and Consolvo, S. (2014). "Experimenting at scale with Google Chrome's SSL warning," *Proceedings of the Conference on Human Factors in Computing Systems: CHI '14*. DOI: 10.1145/2556288.2557292. 121

Fetter, M. and Gross, T. (2011). "PRIMIExperience: Experience sampling via instant messaging," *Proceedings of the Conference on Computer-Supported Cooperative Work: CSCW '11*, pp. 629-32. DOI: 10.1145/1958824.1958931. 87, 159

Fisher, R. A. (1925). *Statistical Methods for Research Workers*. Genesis Publishing Pvt Lt. 102

Fisher, W. W., Piazza, C. C., and Roane, H. S. (Eds.) (2011). *Handbook of Applied Behavior Analysis*. Guilford Press. 151

Fisher, K. and Tucker, J. (2013). "Technical details of time use studies." Oxford, UK: Centre for Time Use Research, University of Oxford. Retrieved from http://www.timeuse.org/information/studies/. 72

Fitts, P. M. (1954). "The information capacity of the human motor system in controlling the amplitude of movement." *Journal of Experimental Psychology*, 47(6), 381-91. DOI: 10.1037/h0055392. 1

Flach, P. (2012). *Machine Learning: The Art and Science of Algorithms that Make Sense of Data*. Cambridge University Press. DOI: 10.1017/cbo9780511973000. 7

Floegel, T. A., Florez-Pregonero, A., Hekler, E. B., and Buman, M. P. (2016). "Validation of consumer-based hip and wrist activity monitors in older adults with varied ambulatory abilities." *The Journals of Gerontology Series A: Biological Sciences and Medical Sciences*, glw098. 16, 37

Fogarty, J., Hudson, S. E., Atkeson, C. G., Avrahami, D., Forlizzi, J., Kiesler, S., Lee, J. C., and Yang, J. (2005). "Predicting human interruptibility with sensors." *ACM Transactions on Computer-Human Interaction (TOCHI)*, 12(1), 119-46. DOI: 10.1145/1057237.1057243. 160

Fogg, B. J. (2009). "A behavior model for persuasive design." *In Proceedings of the 4th International Conference on Persuasive Technology* (p. 40). ACM. DOI: 10.1145/1541948.1541999. 134, 148, 156

Fowler, F. J. (1995). *Improving Survey Questions: Design and Evaluation*. vol. 38. SAGE Publications. 6

Fowler Jr., F. J. (2014). *Survey Research Methods*, 5th ed. SAGE Publications. 6

Freigoun, M., Martin, C. A, Magann, A., Korinek, E., Phatak, S., Rivera, D. E., and Hekler, E. B. (2017). *System Identification of "Just Walk": A Behavioral mHealth Intervention for Promoting Physical Activity*. Manuscript submitted for publication. 35

Froehlich, J., Chen, M. Y., Consolvo, S., Harrison, B., and Landay, J. A. (2007). "MyExperience: A system for in situ tracing and capturing of user feedback on mobile phones," *Proceedings of the International Conference on Mobile Systems, Applications, and Services: MobiSys '07*, pp. 57-70. DOI: 10.1145/1247660.1247670. 87, 98, 159

Froehlich, J., Dillahunt, T., Klasnja, P., Mankoff, J., Consolvo, S., Harrison, B., and Landay, J. A. (2009). "UbiGreen: investigating a mobile tool for tracking and supporting green transportation habits." In *Proceedings of the SIGCHI Conference on Human Factors in Computing Systems* (pp. 1043-52). ACM. DOI: 10.1145/1518701.1518861. 29

Gaonkar, S., Li, J., Choudhury, R. R., Cox, L., and Schmidt, A. (2008). "Micro-blog: sharing and querying content through mobile phones and social participation." In *Proceedings of the 6th International Conference on Mobile Systems, Applications, and Services* (pp. 174-86). ACM. DOI: 10.1145/1378600.1378620. 29

Gaver, B., Boucher, A., Pennington, S., and Walker, B. (2004). "Cultural probes and the value of uncertainty," *Interactions*, pp. 53-6. DOI: 10.1145/1015530.1015555. 78

Gaver, B., Dunne, T., and Pacenti, E. (1999). "Design: Cultural probes," *Interactions*, 6(1), 21-9. DOI: 10.1145/291224.291235. 78

Glanz, K., Rimer, B., and US- N.C.I. (1995). *Theory at a Glance: A Guide for Health Promotion Practice*. NIH. 134

Golbeck, J., Robles, C., Edmondson, M., and Turner, K. (2011). "Predicting personality from twitter." In *Privacy, Security, Risk and Trust (PASSAT) and 2011 IEEE Third International Conference on Social Computing (SocialCom), 2011 IEEE Third International Conference* on (pp. 149-56). IEEE. DOI: 10.1109/passat/socialcom.2011.33. 20

Goldstein, N. J., Cialdini, R. B., and Griskevicius, B. (2008). "A room with a viewpoint: Using social norms to motivate environmental conservation in hotels." *Journal of Consumer Research*, V. 35. DOI: 10.1086/586910. 144

Gould, J. D., Conti, J. and Hovanyecz, T. (1983). "Composing letters with a simulated listening typewriter." *Communications of the ACM*, 26(4), 295-308. DOI: 10.1145/2163.358100. 56

Greenhalgh, T., Stramer, K., Bratan, T., Byrne, E., Mohammad, Y., and Russell, J. (2008). "Introduction of shared electronic records: multi-site case study using diffusion of innovation theory." *BMJ*, 337, a1786. DOI: 10.1136/bmj.a1786. 154

Griswold, W. G., Shanahan, P., Brown, S. W., Boyer, R., Ratto, M., Shapiro, R. B., and Truong, T. M. (2004). ActiveCampus: experiments in community-oriented ubiquitous computing. *Computer*, 37(10), 73-81. 28, 30

Grus, J. (2015). *Data Science from Scratch: First Principles with Python*. O'Reilly Media, Inc. 7

Gubrium, J. F., Holstein, J. A., Marvasti, A. B., and McKinney, K. D. (eds.) (2012). T*he SAGE Handbook of Interview Research: The Complexity of the Craft*. SAGE Publications. 3

Gustafson, D. H., McTavish, F. M., Chih, M. Y., Atwood, A. K., Johnson, R. A., Boyle, M. G., Levy, M. S., Driscoll, H., Chisholm, S. M., Isham, A., and Shah, D. (2014). "A smartphone application to support recovery from alcoholism: a randomized clinical trial." *JAMA Psychiatry*, 71(5), 566-72. DOI: 10.1001/jamapsychiatry.2013.4642. 154

Harboe, G. F., Bentley, F. R., Metcalf, C. J., and Thakkar V. V. (2010). "Method and system for generating a play tree for selecting and playing media content." U.S. Patent 7,685,154, issued March 23, 2010. 50, 51

Harboe, G. and Huang, E. M. (2015). "Real-world affinity diagramming practices: Bridging the paper-digital fap." *Proceedings of the 33rd Annual ACM Conference on Human Factors in Computing Systems.* ACM. DOI: 10.1145/2702123.2702561. 7

Harari, G. M., Gosling, S. D., Wang, R., and Campbell, A. T. (2015). "Capturing situational information with smartphones and mobile sensing methods." *European Journal of Personality*, 29(5), 509-11. DOI: 10.1002/per.2032. 19, 28

Harray, A. J., Boushey, C. J., Pollard, C. M., Delp, E. J., Ahmad, Z., Dhaliwal, S. S., Mukhtar, S. A., and Kerr, D. A. (2015). "A novel dietary assessment method to measure a healthy and sustainable diet using the Mobile Food Record: Protocol and methodology." *Nutrients*, 7(7), 5375-95. DOI: 10.3390/nu7075226. 19

Harris, D. F. (2014). *The Complete Guide to Writing Questionnaires: How to Get Better Information for Better Decisions*. I&M Press. 6

Hekler, E. B., Gardner, C. D., and Robinson, T. N. (2010). "Effects of a course about food and society on college students' eating behaviors." *American Journal of Preventive Medicin*, 38, 543-7. DOI: 10.1016/j.amepre.2010.01.026. 115

Hekler, E. B., Klasnja, P., Froehlich, J. E., and Buman, M. P. (2013a). "Mind the theoretical gap: interpreting, using, and developing behavioral theory in HCI research." In *Proceedings of the SIGCHI Conference on Human Factors in Computing Systems* (pp. 3307-16). ACM. DOI: 10.1145/2470654.2466452. 8, 133, 134, 135, 138, 140, 142, 165

Hekler, E. B., Klasnja, P., Traver, V., and Hendriks, M. (2013b). "Realizing effective behavioral management of health: the metamorphosis of behavioral science methods." *IEEE Pulse*, 4(5), 29-34. DOI: 10.1109/MPUL.2013.2271681. 8, 20, 165

Hekler, E. B., Dubey, G., McDonald, D. Poole, E., Li, V., and Eikey, E. (2014). "Exploring the relationship between changes in weight and utterances in an online weight loss forum." *Journal of Medical Internet Research*, 16(12),e254. DOI: 10.2196/jmir.3735. 33

Hekler, E. B., Buman, M. P., Grieco, L., Rosenberger, M., Haskell, W., and King, A. C. (2015). "Validation of physical activity tracking via android smartphones compared to Actigraph accelerometer: laboratory-based and free-living validation studies." *JMIR mHealth and uHealth*, 3(2),e36. DOI: 10.2196/mhealth.3505. 32, 33

Hekler, E. B., Michie, S., Pavel, M., Rivera, D. E., Collins, L. M., Jimison, H. B., Garnett, C., Parral, S., and Spruijt-Metz, D. (2016a). "Advancing models and theories for digital behavior change interventions." *American Journal of Preventive Medicine*, 51(5), 825-32. DOI: 10.1016/j.amepre.2016.06.013. 8, 103, 161, 164, 165

Hekler, E. B., Klasnja, P., Riley, W. T., Buman, M. P., Huberty, J., Rivera, D. E., and Martin, C. A. (2016b). "Agile science: creating useful products for behavior change in the real world." *Translational Behavioral Medicine*, 6(2),317-28. DOI: 10.1007/s13142-016-0395-7. 165, 166

Hern, A. (2014). "Why Google has 200m reasons to put engineers over designers," *The Guardian*. https://www.theguardian.com/technology/2014/feb/05/why-google-engineers-designers {link verified Dec 28, 2016}. 120

Herrera, J. L., Srinivasan, R., Brownstein, J. S., Galvani, A. P., and Meyers, L. A. (2016). "Disease surveillance on complex social networks." *PLoS Computational Biology*, 12(7), e1004928. DOI: 10.1371/journal.pcbi.1004928. 28

Herzberg, F. (1968). "One more time: how do you motivate employees?" *Harvard Business Review*. The Leader Manager:New York, 433-48. 137

Herzberg, F., Mausner, B., and Snyderman, B. B. (1959). *The Motivation to Work*. John Wiley and Sons. Inc., New York. 137

Holtzmann, H. and Kestner, J. (2017). "Twitter weather." https://www.media.mit.edu/projects/twitter-weather/overview/ (accessed December 26, 2017). 20

Holz, C., Bentley, F., Church, K., and Patel, M. (2015) "I'm just on my phone and they're watching TV: Quantifying mobile device use while watching television." In *Proceedings of ACM TVX*. DOI: 10.1145/2745197.2745210. 39

Hovell, M. F. , Wahlgren, D. R., and Adams, M. A. (2009). "The logical and empirical basis for the Behavioral Ecological Model." In R. J. DiClemente, R. A. Crosby, and M. Kegler (Eds.), *Emerging Theories in Health Promotion Practice and Research: Strategies for Enhancing Public Health* (2nd ed., pp. 415–450). Jossey-Bass:San Francisco. 148

Hsieh, G., Li, I., Dey, A., Forlizzi, J., and Hudson, S. E. (2008). "Using visualizations to increase compliance in experience sampling," *Proceedings of the International Conference on Ubiquitous Computing: UbiComp '08*, pp. 164-7. DOI: 10.1145/1409635.1409657. 159

Hull, B., Bychkovsky, V., Zhang, Y., Chen, K., Goraczko, M., Miu, A., Shih, E., Balakrishnan, H., and Madden, S. (2006). "CarTel: A distributed mobile sensor computing system." In *Proceedings of the 4th International Conference on Embedded Networked Sensor Systems* (pp. 125-38). ACM. DOI: 10.1145/1182807.1182821. 30

Hudson, J. M., Christensen, J., Kellogg, W. A., and Erickson, T. (2002). "I'd Be overwhelmed, but it's just one more thing to do: Availability and interruption in research management," *Proceedings of the Conference on Human Factors in Computing Systems: CHI '02*, pp.97-104. DOI: 10.1145/503376.503394. 84, 85

Hudson, S. E., Fogarty, J., Atkeson, C. G., Avrahami, D., Forlizzi, J., Kiesler, S., Lee, J. C., and Yang, J. (2003). "Predicting human interruptibility with sensors: A Wizard of Oz feasibility study," *Proceedings of the Conference on Human Factors in Computing Systems: CHI '03*, pp. 257-64. DOI: 10.1145/642611.642657. 84, 85, 86

Iachello, G., Truong, K. N., Abowd, G. D., Hayes, G. R., and Stevens, M. (2006). "Prototyping and sampling experience to evaluate ubiquitous computing privacy in the real world," *Proceed-*

ings of the Conference on Human Factors in Computing Systems: CHI '06, pp. 1009-18. DOI: 10.1145/1124772.1124923. 84

Ifenthaler, D. and Schweinbenz, V. (2013). "The acceptance of Tablet-PCs in classroom instruction: The teachers' perspectives." *Computers in Human Behavior*, 29(3), 525-34. DOI: 10.1016/j.chb.2012.11.004. 154

Intille, S., Kukla, C., and Ma, X. (2002). "Eliciting user preferences using image-based experience sampling and reflection," *Proceedings of the Conference on Human Factors in Computing Systems: CHI '02*, pp. 738-9. DOI: 10.1145/506443.506573. 77

Jackson, C., Eliasson, L., Barber, N., and Weinman, J. (2014). "Applying COM-B to medication adherence. A suggested framework for research and interventions." *The European Health Psychologist*, 16(1), 7-17. 155

Jones H. (2010). "Reinforcement-based treatment for pregnant drug abusers (HOME II)" National Institutes of Health, Bethesda, MD. Available at http://clinicaltrials.gov/ct2/show/NCT01177982?term=jones+pregnant&rank=9. 126

Kadushin, C. (2004). *Introduction to Social Network Theory*. Boston, MA. 140

Kahneman, D. and Tversky, A. (1979). "Prospect theory: An analysis of decision under risk." *Econometrica: Journal of the Econometric Society*, 263-91. DOI: 10.2307/1914185. 155

Karkar, R., Zia, J., Vilardaga, R., Mishra, S. R., Fogarty, J., Munson, S. A., and Kientz, J. A. (2015). "A framework for self-experimentation in personalized health." *Journal of the American Medical Informatics Association*, ocv150. 165

Kay, M., Nelson, G., and Hekler, E. (2016). "Researcher-centered design of statistics: Why Bayesian statistics better fit the culture and incentives of HCI." In *Proceedings of SIG-CHI Conference on Human Factors in Computing Systems (CHI'16)*. pp. 4521-32. DOI: 10.1145/2858036.2858465. 118, 164

Khalaf, S. (2015). "Seven years into the mobile revolution: Content is king... again." http://yahoodevelopers.tumblr.com/post/127636051988/seven-years-into-the-mobile-revolution-content-is. 1

King, A. C., Hekler, E. B., Grieco, L. A., Winter, S. J., Sheats, J. L., Buman, M. P., Banerjee, B., Robinson, T. N., and Cirimele, J. (2013). "Harnessing different motivational frames via mobile phones to promote daily physical activity and reduce sedentary behavior in aging adults," *PLoS One*, 8(4), e62613. DOI: 10.1371/journal.pone.0062613. 77, 150, 151

King, A. C., Hekler, E. B., Grieco, L. A., Winter, S. J., Sheats, J. L., Buman, M. P., Banerjee, B., Robinson, T. N., and Cirimele, J. (2016). "Effects of three motivationally targeted mobile device applications on initial physical activity and sedentary behavior change in midlife and older adults: A randomized trial," *PLoS One*, 11(6), e0156370. DOI: 10.1371/journal. pone.0156370. 29, 77, 150

King, R. and Churchill, E. F. (2017). *Designing with Data: Improving User Experience with Large Scale User Testing*. O'Reilly Media. 162

Klasnja, P., Harrison, B. L., LeGrand, L., LaMarca, A., Froehlich, J., and Hudson, S. E. (2008). "Using wearable sensors and real time inference to understand human recall of routine activities," *Proceedings of the 10th International Conference on Ubiquitous Computing: Ubi-Comp '08*, pp. 154-63. DOI: 10.1145/1409635.1409656. 78

Klasnja, P., Consolvo, S., Jung, J., Greenstein, B. M., LeGrand, L., Powledge, P., and Wetherall, D. (2009). "When I am on Wi-Fi, I am fearless": Privacy concerns and practices in everyday Wi-Fi use," *Proceedings of the Conference on Human Factors in Computing Systems: CHI '09*, pp. 1993-2002. DOI: 10.1145/1518701.1519004. 92

Klasnja, P., Consolvo, S., and Pratt, W. (2011). "How to evaluate technologies for health behavior change in HCI research. "In *Proceedings of the SIGCHI Conference on Human Factors in Computing Systems* (pp. 3063-72). ACM. DOI: 10.1145/1978942.1979396. 140, 153

Klasnja, P., Hekler, E. B., Shiffman, S., Boruvka, A., Almirall, D., Tewari, A., and Murphy, S. A. (2015). "Microrandomized trials: An experimental design for developing just-in-time adaptive interventions." *Health Psychology*, 34(S), 1220. DOI: 10.1037/hea0000305. 128, 164

Klasnja, P., Hekler, E.B., Korinek, E.V., Harlow, J., and Mishra, S. (2017) "Toward usable evidence: Optimizing knowledge accumulation in HCI research on health behavior." *Proceedings of the SIGCHI Conference on Human Factors in Computing Systems* (CHI '17). 131, 164, 165

Klotzbach, C. (2016). "Enter the matrix: App retention and engagement." http://flurrymobile. tumblr.com/post/144245637325/appmatrix. 25

Kravitz R. L., Duan N., eds, and the DEcIDE Methods Center N-of-1 Guidance Panel (Duan, N., Eslick, I., Gabler, N. B., Kaplan, H. C., Kravitz, R. L., Larson, E. B., Pace, W. D., Schmid, C. H., Sim, I., and Vohra, S.). (2014). "Design and implementation of N-of-1 trials: A user's guide." *AHRQ Publication No. 13(14)-EHC122-EF*. Agency for Healthcare Research and Quality:Rockville, MD. 111, 112, 165

Kuhn, T. S. (2012). *The Structure of Scientific Revolutions*. University of Chicago press. DOI: 10.7208/chicago/9780226458144.001.0001. 102

Landauer, T. K. (1987). "Psychology as a mother of invention." *Proceedings of the SIGCHI/GI Conference on Human Factors in Computing Systems and Graphics Interface: SIGCHI/GI '87*, pp. 333-5. DOI: 10.1145/29933.275653. 56

Lathia, N., Rachuri, K. K., Mascolo, C., and Rentfrow, P. J. (2013). "Contextual dissonance: Design bias in sensor-based experience sampling methods," *Proceedings of the International Conference on Ubiquitous Computing: UbiComp '13*, pp. 183-92. DOI: 10.1145/2493432.2493452. 159

Lazarus, R. S., and Folkman, S. (1987). "Transactional theory and research on emotions and coping." *European Journal of Personality*, 1(3), 141-69. DOI: 10.1002/per.2410010304. 149

Lazer, D., Pentland, A. S., Adamic, L., Aral, S., Barabasi, A. L., Brewer, D., Christakis, N., Contractor, N., Fowler, F., Gutmann, M., Jebara, T., King, G., Macy, M, Roy, D., and Van Alstyne, M. (2009). "Life in the network: the coming age of computational social science." *Science*, 323(5915), 721. DOI: 10.1126/science.1167742. 20

Lee, J., Walker, E., Kay, M., Burleson, W., Buman, M. P., and Hekler, E. B. (2017). "Self-experimentation for behavior change: Design and formative evaluation of two approaches." *Proceedings of the SIGCHI Conference on Human Factors in Computing Systems* (CHI '17). 119, 165

Lee, K., Flinn, J., and Noble, B. (2015). "The case for operating system management of user attention." In P*roceedings of the 16th International Workshop on Mobile Computing Systems and Applications* (pp. 111-6). ACM. DOI: 10.1145/2699343.2699362. 160

Lei, H., Nahum-Shani, I., Lynch, K., Oslin, D., and Murphy, S. A. (2012). "A "SMART" design for building individualized treatment sequences." *Annual Review of Clinical Psychology*, 8. DOI: 10.1146/annurev-clinpsy-032511-143152. 126, 127

Leveson, N. G. and Turner, C. S. (1993). "An investigation of the Therac-25 accidents," *Computer*, 26(7), 18-41. DOI: 10.1109/MC.1993.274940. 42

Ljung, L. (1998). "System identification." In *Signal Analysis and Prediction* (pp. 163-73). Birkhäuser Boston. DOI: 10.1007/978-1-4612-1768-8_11. 7

Locke, E. A. and Latham, G. P. (2002). "Building a practically useful theory of goal setting and task motivation: A 35-year odyssey." *American Psychologist*, 57(9), 705. DOI: 10.1037/0003-066X.57.9.705. 134, 152, 154

Loock, C. M., Staake, T., and Thiesse, F. (2013). "Motivating energy-efficient behavior with green IS: An investigation of goal setting and the role of defaults." *MIS Quarterly*, 37(4), 1313-32. 154

Lu, H., Frauendorfer, D., Rabbi, M., Mast, M.S., Chittaranjan, G.T., Campbell, A.T., Gatica-Perez, D., and Choudhury, T. (2012). "StressSense: Detecting stress in unconstrained acoustic environments using smartphones," *Proceedings of the 2012 ACM Conference on Ubiquitous Computing: UbiComp '12*, pp. 351-60. DOI: 10.1145/2370216.2370270. 31

Lu, H., Pan, W., Lane, N. D., Choudhury, T., and Campbell, A. T. (2009). "SoundSense: scalable sound sensing for people-centric applications on mobile phones." In *Proceedings of the 7th International Conference on Mobile Systems, Applications, and Services* (pp. 165-78). ACM. DOI: 10.1145/1555816.1555834. 19

Luarn, P. and Lin, H. H. (2005). "Toward an understanding of the behavioral intention to use mobile banking." *Computers in Human Behavior*, 21(6), 873-91. DOI: 10.1016/j.chb.2004.03.003. 139

Madan, A., Cebrian, M., Lazer, D., and Pentland, A. (2010). "Social sensing for epidemiological behavior change." In *Proceedings of the 12th ACM International Conference on Ubiquitous Computing* (pp. 291-300). ACM. DOI: 10.1145/1864349.1864394. 19

Mandel, E. (2014). "How the Napa earthquake affected Bay Area sleepers," *The Jawbone Blog*, https://jawbone.com/blog/napa-earthquake-effect-on-sleep/ {link verified Dec 26, 2016} 28

Mankoff, J., Dey, A. K., Hsieh, G., Kientz, J., Lederer, S., and Ames, M. (2003). "Heuristic evaluation of ambient displays," *Proceedings of the Conference on Human Factors in Computing Systems: CHI '03*, pp. 169-76. DOI: 10.1145/642611.642642. 60

Martella, C., Miraglia, A., Frost, J., Cattani, M., and van Steen, M. (2016). "Visualizing, clustering, and predicting the behavior of museum visitors." *Pervasive and Mobile Computing*. DOI: 10.1016/j.pmcj.2016.08.011. 15

Martin, C. K., Correa, J. B., Han, H., Allen, H. R., Rood, J. C., Champagne, C. M., Gunturk, B. K., and Bray, G. A. (2012). "Validity of the remote food photography method (RFPM) for estimating energy and nutrient intake in near real-time." *Obesity*, 20(4), 891-9. DOI: 10.1038/oby.2011.344. 19

Martín, C. A., Hekler, E. B., and Rivera, D. E. (2015). "Design of informative identification experiments for behavioral interventions, *17th IFAC Symposium on System Identification (SYSID 2015)* via IFAC-PapersOnLine, 48(28), 1325-30. DOI: 10.1016/j.ifacol.2015.12.315. 129, 130

Martín, C. A., Rivera, D. E., and Hekler, E. B. (2016). "A decision framework for an adaptive behavioral intervention for physical activity using hybrid model predictive control." Paper presented at the *2016 American Control Conference (ACC)*, 3576-3581. DOI: 10.1109/ACC.2016.7525468. 129, 130

Mast, M. S., Gatica-Perez, D., Frauendorfer, D., Nguyen, L., and Choudhury, T. (2015). "Social sensing for psychology automated interpersonal behavior assessment." *Current Directions in Psychological Science*, 24(2), 154-60. DOI: 10.1177/0963721414560811. 21, 28

Matthews, T., Liao, K., Turner, A., Berkovich, M., Reeder, R., and Consolvo, S. (2016). " 'She'll just grab any device that's closer': A study of everyday device and account sharing in households," *Proceedings of the ACM Conference on Human Factors in Computing Systems: CHI '16*, pp. 5921-32. DOI: 10.1145/2858036.2858051. 12, 31, 38, 99

Matthews, T., O'Leary, K., Turner, A., Sleeper, M., Woelfer, J.P., Shelton, M., Manthorne, C., Churchill, E.F., and Consolvo, S. (2017). "Stories from survivors: Privacy & security practices when coping with intimate partner abuse." *Proceedings of the ACM Conference on Human Factors in Computing Systems: CHI '17.* 12

McGarraugh, G. (2009). "The chemistry of commercial continuous glucose monitors." *Diabetes Technology and Therapeutics*, 11(S1), S-17. DOI: 10.1089/dia.2008.0133. 22

McGregor, M., Brown, B., and McMillan, D. (2014). "100 days of iPhone use: mobile recording in the wild." In *CHI'14 Extended Abstracts on Human Factors in Computing Systems*, pp. 2335-40. DOI: 10.1145/2559206.2581296. 77

Mehrotra, A., Hendley, R., and Musolesi, M. (2016). "PrefMiner: mining user's preferences for intelligent mobile notification management." In *Proceedings of the 2016 ACM International Joint Conference on Pervasive and Ubiquitous Computing* (pp. 1223-1234). ACM. DOI: 10.1145/2971648.2971747. 160

Mehrotra, A., Musolesi, M., Hendley, R., and Pejovic, V. (2015). "Designing content-driven intelligent notification mechanisms for mobile applications." *Proceedings of the 2015 ACM International Joint Conference on Pervasive and Ubiquitous Computing: UbiComp '15.* DOI: 10.1145/2750858.2807544. 160

Meschtscherjakov, A., Trösterer, S., Döttlinger, C., Wilfinger, D., and Tscheligi, M. (2013). "Computerized experience sampling in the car -- Issues and challenges," *Proceedings of the International Conference on Automotive User Interfaces and Interactive Vehicular Applications: AutomotiveUI '13*, pp. 220-3. DOI: 10.1145/2516540.2516565. 84

Meschtscherjakov, A., Wilfinger, D., Osswald, S., Perterer, N., and Tscheligi, M. (2012). "Trip experience sampling: Assessing driver experience in the field," *Proceedings of the International Conference on Automotive User Interfaces and Interactive Vehicular Applications: AutomotiveUI '12*, pp. 225-32. DOI: 10.1145/2390256.2390294. 84

Metcalf, C. J. and Harboe, G. (2006). "Sunday is family day," *Ethnographic Praxis in Industry Conference Proceedings*, 2006(1), 49-59. DOI: 10.1111/j.1559-8918.2006.tb00035.x. 46

Michie, S. F., West, R., Campbell, R., Brown, J., and Gainforth, H. (2014). *ABC of Behaviour Change Theories*. Silverback Publishing. 146, 147

Michie, S., van Stralen, M. M., and West, R. (2011). "The behaviour change wheel: a new method for characterising and designing behaviour change interventions." *Implementation Science*, 6(1), 42. DOI: 10.1186/1748-5908-6-42. 136, 155

Miller, W. R. and Rollnick, S. (2012). *Motivational Interviewing: Helping People Change*. Guilford Press. 146

Mohr, D. C., Cheung, K., Schueller, S. M., Brown, C. H., and Duan, N. (2013). "Continuous evaluation of evolving behavioral intervention technologies." *American Journal of Preventive Medicine*, 45(4), 517-23. DOI: 10.1016/j.amepre.2013.06.006. 129

Mohr, D. C., Ho, J. Hart, T. L., Baron, K. G., Berendsen, M., Beckner, V., Cai, X., Cuijpers, P., Spring, B., Kinsinger, S. W., Schroder, K. E., and Duffecy, J. (2014). "Control condition design and implementation features in controlled trials: A meta-analysis of trials evaluating psychotherapy for depression," *Translational Behavioral Medicine*, 4, 407-23. DOI: 10.1007/s13142-014-0262-3. 118, 131

Moore, G. A. (1991). *Crossing the Chasm: Marketing and Selling Technology Products to Mainstream Consumers*. Harpers Business, New York. 139

Murphy S. A. (2005). "An experimental design for the development of adaptive treatment strategies." *Statistics in Medicine*, 24,1455–81. DOI: 10.1002/sim.2022. 125

Mynatt, E. D., Rowan, J., Jacobs, A., and Craighill, S. (2001). "Digital family portraits: Supporting peace of mind for extended family members," *Proceedings of the Conference on Human Factors in Computing Systems: CHI '01*, pp. 333-40. DOI: 10.1145/365024.365126. 57

Nahum-Shani, I., Hekler, E. B., and Spruijt-Metz, D. (2015). "Building health behavior models to guide the development of just-in-time adaptive interventions: A pragmatic framework." *Health Psychology*, 34(S), 1209. DOI: 10.1037/hea0000306. 8, 160, 165

Nezlek, J. B., Wheeler, L., and Reis, H. T. (1983). "Studies of social participation," *New Directions for Methodology of Social and Behavioral Science*, 15, pp. 57-73. 84

Nielsen, J. (1993). *Usability Engineering*. Morgan Kaufmann Academic Press, San Diego, CA. 4, 50, 61

Nielsen, J. (2012). "Usability 101: Introduction to usability," Nielsen Norman Group website. http://www.nngroup.com/articles/usability-101-introduction-to-usability/ {link verified Nov 27, 2016} 4, 50

Nielsen, J. (2012). "Thinking aloud: The #1 usability tool," Nielsen Norman Group website. https://www.nngroup.com/articles/thinking-aloud-the-1-usability-tool/ {link verified Nov 27, 2016}. 61

Nielsen, J. and Mack, R. L. (1994). *Usability Inspection Methods*, John Wiley and Sons, Inc. DOI: 10.1145/259963.260531. 1, 4

Norman, D.A. (1981). "Categorization of action slips." *Psychology Review*, 88(1), pp. 1-15. DOI: 10.1037/0033-295X.88.1.1. 72

Nundy, S., Dick, J. J., Chou, C. H., Nocon, R. S., Chin, M. H., and Peek, M. E. (2014). "Mobile phone diabetes project led to improved glycemic control and net savings for Chicago plan participants." *Health Affairs*, 33(2), 265-72. DOI: 10.1377/hlthaff.2013.0589. 116

O'Connell, C. (2016). "23% of users abandon an app after one use." http://info.localytics.com/blog/23-of-users-abandon-an-app-after-one-use. 25

Palen, L. and Salzman, M. (2002). "Voice-mail diary studies for naturalistic data capture under mobile conditions," *Proceedings of the Conference on Computer-Supported Cooperative Work: CSCW '02*. pp. 87-95. DOI: 10.1145/587078.587092. 72

Patrick, K., Marshall, S. J., Davila, E. P., Kolodziejczyk, J. K., Fowler, J. H., Calfas, K. J., Huang, J. S., Rock, C. L., Griswold, W. G., Gupta, A., Merchant, G., Norman, G. J., Raab, F., Donohue, M. C., Fogg, B. J., and Robinson, T. N. (2014). "Design and implementation of a randomized controlled social and mobile weight loss trial for young adults (project SMART)." *Contemporary Clinical Trials*, 37(1), 10-18. DOI: 10.1016/j.cct.2013.11.001. 141, 155

Patrick, K., Hekler, E. B., Estrin, D., Mohr, D. C., Riper, H., Crane, D., Godino, J., and Riley, W. T. (2016). "The pace of technologic change: Implications for digital health behavior intervention research." *American Journal of Preventive Medicine*, 51(5), 816-24. DOI: 10.1016/j.amepre.2016.05.001. 8, 165

Pearl, J., Glymour, M., and Jewell, N. P. (2016). *Causal Inference in Statistics: A Primer*. John Wiley and Sons. 7

Pentland, A. (2014). *Social Physics: How Good Ideas Spread- The Lessons from a New Science*. Penguin. 20, 28

Perera, C., Zaslavsky, A., Christen, P. and Georgakopoulos, D., (2014). "Context aware computing for the internet of things: A survey." *IEEE Communications Surveys and Tutorials*, 16(1), 414-454. DOI: 10.1109/SURV.2013.042313.00197. 21

Pielot, M., Church, K., and De Oliveira, R. (2014) "An in-situ study of mobile phone notifications." In *Proceedings of the 16th International Conference on Human-Computer Interaction with Mobile Devices and Services*, pp. 233-242. ACM. DOI: 10.1145/2628363.2628364. 27, 160

Pierdicca, R., Liciotti, D., Contigiani, M., Frontoni, E., Mancini, A., and Zingaretti, P. (June). "Low cost embedded system for increasing retail environment intelligence." In *Multimedia and Expo Workshops (ICMEW), 2015 IEEE International Conference on* (pp. 1-6). IEEE. DOI: 10.1109/icmew.2015.7169771. 15

Popper, K. (2005). *The Logic of Scientific Discovery.* Routledge. 102

Prochaska, J. O. and DiClemente, C. C. (1994). *The Transtheoretical Approach: Crossing Traditional Boundaries of Therapy.* Krieger Pub. 134, 146

Prochaska, J. O. and Velicer, W. F. (1997). "The transtheoretical model of health behavior change." *American Journal of Health Promotion*, 12(1), 38-48. DOI: 10.4278/0890-1171-12.1.38. 138, 155

Prochaska, J. O., Wright, J. A., and Velicer, W. F. (2008). "Evaluating theories of health behavior change: A hierarchy of criteria applied to the transtheoretical model." *Applied Psychology-International Association of Applied Psychology,* 57(4),561-88. DOI: 10.1111/j.1464-0597.2008.00345.x. 138

Rabbi, M., Aung, M. H., Zhang, M., and Choudhury, T. (2015). "MyBehavior: automatic personalized health feedback from user behaviors and preferences using smartphones." In *Proceedings of the 2015 ACM International Joint Conference on Pervasive and Ubiquitous Computing* (pp. 707-718). ACM. DOI: 10.1145/2750858.2805840. 29, 113, 156, 161

Raiff, B. R. and Dallery, J. (2010). "Internet-based contingency management to improve adherence with blood glucose testing recommendations for teens with type I diabetes." *Journal of Applied Behavior Analysis*, 43(3), 487-91. DOI: 10.1901/jaba.2010.43-487. 112

Reder, S., Ambler, G., Philipose, M., and Hedrick, S. (2010). "Technology and Long-term Care (TLC): A pilot evaluation of remote monitoring of elders," *Gerontechnology*, 9(1), 18-31. DOI: 10.4017/gt.2010.09.01.002.00. 60

Reeves, M.P. (1979) (first published 1913). *Round About a Pound a Week.* Virago:London. 72

Resnick, P. and Varian, H. R. (1997). "Recommender systems." *Communications of the ACM* 40(3), 56-8. DOI: 10.1145/245108.245121. 20

Rettig, M. (1994). "Prototyping for tiny fingers," Communications of the ACM, 37(4), 21-7. DOI: 10.1145/175276.175288. 61

Riley, W. T., Rivera, D. E., Atienza, A. A., Nilsen, W., Allison, S. M., and Mermelstein, R. (2011). "Health behavior models in the age of mobile interventions: are our theories up to the task?" *Translational Behavioral Medicine*, 1(1), 53-71. DOI: 10.1007/s13142-011-0021-7. 165

Robinson, E., Aveyard, P., Daley, A., Jolly, K., Lewis, A., Lycett, D., and Higgs, S. (2013a). "Eating attentively: a systematic review and meta-analysis of the effect of food intake memory and awareness on eating." *The American journal of clinical nutrition*, ajcn-045245. 137

Robinson, E., Higgs, S., Daley, A. J., Jolly, K., Lycett, D., Lewis, A., and Aveyard, P. (2013b). "Development and feasibility testing of a smartphone based attentive eating intervention." *BMC Public Health*, 13(1), 639. DOI: 10.1186/1471-2458-13-639. 140

Rogers, E. (1962). *Diffusion of innovations*. Free Press of Glencoe:New York, 1, 79-134. 139, 154

Roessler, P., Consolvo, S., and Shelton, B. (2004). "Phase #2 of Computer-Supported Coordinated Care Project." Intel Research Seattle Tech Report IRS-TR-04-006. 58

Rubin, J. (1994). *Handbook of Usability Testing: How to Plan, Design, and Conduct Effective Tests*. John Wiley and Sons, Inc. 60

Rubin, J. and Chisnell, D. (2008). *Handbook of Usability Testing: How to Plan, Design and Conduct Effective Tests*. 2nd ed. John Wiley and Sons. 5

Ryan, R. M., and Deci, E. L. (2000). "Self-determination theory and the facilitation of intrinsic motivation, social development, and well-being." *American Psychologist*, 55(1), 68. DOI: 10.1037/0003-066X.55.1.68. 146, 154

Santillana, M., Nguyen, A. T., Dredze, M., Paul, M. J., Nsoesie, E. O., and Brownstein, J. S. (2015). "Combining search, social media, and traditional data sources to improve influenza surveillance." *PLoS Computational Biology*, 11(10), e1004513. DOI: 10.1371/journal.pcbi.1004513. 28

Scholtz, J. (2004). "Usability evaluation." National Institute of Standards and Technology. 1

Scholtz, J. and Consolvo, S. (2004). "Toward a framework for evaluating ubiquitous computing applications." *IEEE Pervasive Computing*, 3(2), 82-8. DOI: 10.1109/MPRV.2004.1316826. 2

Sellen, A. J. (1994). "Detection of everyday errors." *Applied Psychology: An International Review*, 43(4), 475-98. DOI: 10.1111/j.1464-0597.1994.tb00841.x. 72

Shadish, W. R., Cook, T. D., and Campbell, D. T. (2002). *Experimental and Quasi-experimental Designs for Generalized Causal Inference*. Houghton, Mifflin and Company. 102, 111, 112, 116, 119

Shirazi, A.S., Henze, N., Dingler, T., Pielot, M., Weber, D., and Schmidt, A. (2014). Large-scale assessment of mobile notifications. In *Proceedings of the SIGCHI Conference on Human Factors in Computing Systems*: CHI '14. DOI: 10.1145/2556288.2557189. 163

Simon, H. A. (1956). "Rational choice and the structure of the environment," *Psychological Review*, 63(2), 129-38. DOI: 10.1037/h0042769. 51

Skinner, B. F. (1953). *Science and Human Behavior*. New York: Free Press.151

Skinner, B. F. (1981). Selection by consequences. *Science*, 213, 501–4. DOI: 10.1126/science.7244649. 151

Smith, A. (2015). US smartphone use in 2015. Pew Research Center, 1. 32

Spence, A. and Pidgeon, N. (2010). "Framing and communicating climate change: The effects of distance and outcome frame manipulations." *Global Environmental Change*, 20(4), 656-67. DOI: 10.1016/j.gloenvcha.2010.07.002. 155

Spruijt-Metz, D., Hekler, E., Saranummi, N., Intille, S., Korhonen, I., Nilsen, W., Rivera, D. E., Spring, B., Michie, S., Asch, D. A., Sanna, A., Salcedo, V. T, Kukakfa, R., and Pavel, M. (2015). "Building new computational models to support health behavior change and maintenance: new opportunities in behavioral research." *Translational Behavioral Medicine*, 5(3), 335-346. DOI: 10.1007/s13142-015-0324-1. 8, 165

Stone, A. A. and Shiffman, S. (1994). "Ecological momentary assessment (EMA) in behavioral medicine." *Annals of Behavioral Medicine*, 16, pp.199-202.83

Strauss, A. and Corbin, J. (1998). *Basics of Qualitative Research: Techniques and Procedures for Developing Grounded Theory*. Sage Publications, Inc. 7

Strauss, A. and Corbin, J. (1990). *Basics of Qualitative Research: Grounded Theory Procedures and Techniques. 2nd ed.* Sage Publications, Inc. 9

Strecher, V. J., McClure, J. B., Alexander, G. L., Chakraborty, B., Nair, V. N., Konkel, J. M., Greene, S. M., Collins, L. M., Carlier, C. C., Wiese, C. J., Little, R. J., Pomerleau, C. S., and Pomerleau, O. F. (2008). Web-based smoking-cessation programs: results of a randomized trial. *American Journal of Preventive Medicine*, 34(5), 373-381. DOI: 10.1016/j.amepre.2007.12.024. 123, 124, 125

Tanner, T. (2013). "The problem of alarm fatigue." *Nursing for Women's Health*. 17(2):153-157. DOI: 10.1111/1751-486X.12025. 160

Tausczik, Y. R. and Pennebaker, J. W. (2010). The psychological meaning of words: LIWC and computerized text analysis methods. *Journal of Language and Social Psychology*, 29, 24-54. DOI: 10.1177/0261927X09351676. 20, 32, 34

Thompson, C., Johnson, M., Egelman, S., Wagner, D., and King, J. (2013). "When It's Better to Ask Forgiveness than Get Permission: Attribution Mechanisms for Smartphone

Resources." *Proceedings of the 9th Symposium on Usable Privacy and Security: SOUPS '13*, Newcastle, UK. DOI: 10.1145/2501604.2501605. 60

Turgeman, Y., Alm E., and Ratti, C. (2014) "Smart toilets and sewers are coming." *Wired UK*. http://senseable.mit.edu.ezproxy1.lib.asu.edu/papers/pdf/20140321_Turgeman_etal_SmartToilets_Wired.pdf. 21

Umstattd, M. R., Wilcox, S., Saunders, R., Watkins, K., and Dowda, M. (2008). "Self-regulation and physical activity: The relationship in older adults." *American Journal of Health Behavior*, 32(2), 115-24. DOI: 10.5993/AJHB.32.2.1. 150

van Duyne, D., Landay, J.A., and Hong, J.I. (2007). *The Design of Sites: Patterns for Creating Winning Web Sites*. Prentice Hall, 2nd Ed. 1

Venkatesh, V., Morris, M. G., Davis, G. B., and Davis, F. D. (2003). "User acceptance of information technology: Toward a unified view." *MIS Quarterly*, 425-78. 139, 154

Vijayasarathy, L. R. (2004). "Predicting consumer intentions to use on-line shopping: the case for an augmented technology acceptance model." *Information and Management*, 41(6), 747-62. DOI: 10.1016/j.im.2003.08.011. 139

Voit, A., Machulla, T., Weber, D., Schwind, V., Schneegass, S., and Henze, N. (2016). "Exploring notifications in smart home environments." Paper presented at: *Proceedings of the 18th International Conference on Human-Computer Interaction with Mobile Devices and Services Adjunct 2016*. DOI: 10.1145/2957265.2962661. 160

Want, R., Pering, T., Danneels G., Kumar, M., Sundar, M., and Light, J. (2002). "The personal server: Changing the way we think about ubiquitous computing." *Proceedings of the International Conference on Ubiquitous Computing: UbiComp '02*, pp. 194-209. DOI: 10.1007/3-540-45809-3_15. 86

Watkins, C. (2013). "Formspring - A Postmortem." http://blog.capwatkins.com/formspring-a-postmortem {link verified Dec 19, 2016}. 41

Weber, D., Voit, A., Le, H. V., and Henze, N. (2016). "Notification dashboard: enabling reflection on mobile notifications." Paper presented at: *Proceedings of the 18th International Conference on Human-Computer Interaction with Mobile Devices and Services Adjunct 2016*. DOI: 10.1145/2957265.2962660. 160

Weiser, M. (1991). "The computer for the 21st century." *Scientific American*, 265(3), 66-75. DOI: 10.1038/scientificamerican0991-94. 2

Weppner, J., Lukowicz, P., Serino, S., Cipresso, P., Gaggioli, A., and Riva, G. (2013). "Smartphone based experience sampling of stress-related events," *Proceedings of the International Con-*

ference on Pervasive Computing Technologies for Healthcare and Workshops, pp. 464-67. DOI: 10.4108/icst.pervasivehealth.2013.252358. 84

Wheeler, L. and Reis, H. (1991). "Self-recording of everyday life events: Origins, types, and uses." *Journal of Personality*, 59, 339-54. DOI: 10.1111/j.1467-6494.1991.tb00252.x. 84

Wilson, J. and Rosenberg, D. (1988). "Rapid prototyping for user interface design." In M. Helander (ed.), *Handbook of* 139-*Computer Interaction*, (pp. 859-875). New York, North-Holland. DOI: 10.1016/B978-0-444-70536-5.50044-0. 55

Wu, I. L., Li, J. Y., and Fu, C. Y. (2011). "The adoption of mobile healthcare by hospital's professionals: an integrative perspective." *Decision Support Systems*, 51(3), 587-96. DOI: 10.1016/j.dss.2011.03.003. 139

Wyatt, D., Choudhury, T., Bilmes, J., and Kitts, J. A. (2011). Inferring colocation and conversation networks from privacy-sensitive audio with implications for computational social science. *ACM Transactions on Intelligent Systems and Technology (TIST)*, 2(1), 7. DOI: 10.1145/1889681.1889688. 19, 35

Yeh, T. and Darrell, T. (2007). *Photo-Oriented Question— A Multi-Modal Approach to Information Retrieval*. MIT CSAIL Research Abstracts, 2007. 56

Author Biographies

Sunny Consolvo, Ph.D., CIPP/US
User Experience Researcher
Google

sconsolvo@google.com
1600 Amphitheatre Parkway
Mountain View, CA 94043

Sunny Consolvo leads Google's Security & Privacy User Experience team. Sunny and her team focus on usable privacy and security (e.g., understanding how people share their mobile devices). Sunny previously worked as a Research Scientist at Intel Labs Seattle where she investigated how to use mobile technologies to encourage health & wellness and to help people be more aware of the privacy implications of sensing and inference systems. She has also designed and evaluated tools to help people be more aware of what they expose when they use Wi-Fi, examined privacy implications of location-enhanced technologies, and developed technologies to help elders age in place. Sunny received her Ph.D. in Information Science from the University of Washington. She is a member of the Editorial Board for IEEE Pervasive Computing and the PACM on Interactive, Mobile, Wearable, and Ubiquitous Technologies. She became a Certified Information Privacy Professional (US) in 2013.

Frank R. Bentley
Senior Principal Researcher
Yahoo

fbentley@yahoo-inc.com
701 1st Avenue, Building D
Sunnyvale, CA 94089

Frank Bentley is a Senior Principal Researcher at Yahoo, where he leads User Research for Yahoo Mail, Messenger, Flickr, and View. Frank's research identifies new product opportunities in communication and digital media based on findings from iterative user research and through the rapid prototyping and field evaluation of new concepts. For the past 12 years, he has taught a Mobile HCI class at MIT, which has also reached over 75,000 students on the EdX platform, and is currently teaching a new course, Understanding Users, at Stanford. Frank's first book, *Building Mobile Experiences*, focuses on designing for the novel opportunities that mobility provides by utilizing ethnographic research throughout the design and product development process.

Eric B. Hekler, Ph.D.
Assistant Professor, School of Nutrition & Health Promotion
Arizona State University

ehekler@asu.edu
602-827-2271
500 North 3rd St., Room 121
Phoenix, AZ 85004
http://www.designinghealth.org/

Eric Hekler directs the Designing Health Lab at Arizona State University. His research focuses on individualized and "precise" behavior change for long-term health via digital health technologies and he is developing a research process called Agile Science. Examples of his work include NSF-funded research focused on developing mathematical models for guiding an intervention that determines an individualized "ambitious but doable" daily step goal to strive for each day. His Google-funded work focused on teaching individuals the fundamentals of behavior change and self-experimentation and giving them tools (e.g., home sensors and feedback) to allow them to self-experiment with behavior change techniques to optimize their health. Prior to ASU, Dr. Hekler completed his postdoctoral training at Stanford University and received his Ph.D. in Clinical Health Psychology from Rutgers University.

Sayali S. Phatak
Ph.D. Student
School of Nutrition and Health Promotion
Arizona State University

sayali.phatak@asu.edu
500 N 3rd St
Phoenix, AZ 85004

Sayali Phatak is a Ph.D. student in the School of Nutrition and Health Promotion at Arizona State University. She works at the Designing Health Lab with Dr. Eric Hekler. Her research revolves around the design and development of personalized behavioral interventions using digital technologies. Over the last two years, she has worked on an NSF-funded project that uses a dynamical systems modeling approach to inform an app that can assign personalized daily step goals based on what it knows about the user. Her dissertation work is focused on systematizing self-experimentation research, particularly, developing a tool that can assist citizen scientists in designing data-collection protocols that are appropriate for their research questions.

Printed in the United States
by Baker & Taylor Publisher Services